THE
Calculus
Wars

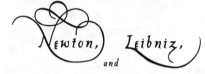
Newton, Leibniz, and

THE GREATEST MATHEMATICAL CLASH OF ALL TIME

Jason Socrates Bardi

THUNDER'S MOUTH PRESS
NEW YORK

The Calculus Wars
NEWTON, LEIBNIZ, AND THE GREATEST MATHEMATICAL CLASH OF ALL TIME

Published by
Thunder's Mouth Press
An imprint of Avalon Publishing Group, Inc.
245 West 17th Street, 11th Floor
New York, NY 10011

AVALON

Copyright © 2006 by Jason Socrates Bardi

First printing May 2006

All rights reserved. No part of this publication may be reproduced or transmitted
in any form or by any means, electronic or mechanical, including photocopy,
recording, or any information storage and retrieval system now known or to be
invented, without permission in writing from the publisher, except by a reviewer
who wishes to quote brief passages in connection with a review written for
inclusion in a magazine, newspaper, or broadcast.

Library of Congress Cataloging-in-Publication Data is available.

ISBN: 1-56025-706-7
ISBN 13: 978-1-56025-706-6

9 8 7 6 5 4 3 2 1

Book design by Pauline Neuwirth, Neuwirth & Associates, Inc.

Printed in the United States of America
Distributed by Publishers Group West

Contents

Preface

AT THE BEGINNING of the eighteenth century, Gottfried Wilhelm Leibniz (1646–1716) and Sir Isaac Newton (1642–1726) were about to go to war.

For more than ten bitter years, these two brilliant figures in German and British mathematics would fight a brutal public battle to the ends of their lives, in which each man defended his own right to claim intellectual ownership of calculus—the branch of mathematical analysis useful for investigating everything from geometrical shapes to the orbits of planets in motion around the sun.

One of the greatest intellectual legacies of the seventeenth century, calculus was developed first by Newton in his creative years of 1665 and 1666, when he was a young Cambridge University student on retreat in his country estate. Suddenly cut off from his professors and classmates, Newton spent two years in near-absolute isolation doing experiments and thinking about the physical laws that govern the universe. What emerged from these years is perhaps the greatest single body of knowledge any scientist has ever produced in such a short period. Newton made major discoveries concerning modern optics, fluid mechanics, the physics of tides, the laws or motion, and the theory of universal gravitation, to name a few.

Most important, Newton had invented calculus, which he called his method of fluxions and fluents. But he kept this work a closely held secret for most of his life. He preferred to circulate private copies of his projects among his friends, and did not publish any of his calculus work until decades after its inception.

Leibniz came upon calculus ten years later, during the prolific time he spent in Paris around 1675. Over the next ten years, he refined his discovery and developed a completely original system of symbols and notations. Though second in time, he was first to publish his calculus, which he did in two papers that appeared in 1684 and 1686. With these two papers, Leibniz was able to claim intellectual ownership for his original invention of calculus. And calculus was such a promising invention that, by the year 1700, Leibniz would be regarded by many in Europe as one of the greatest mathematicians alive.

Leibniz and Newton both had a claim of ownership on calculus, and today they are generally regarded as twin independent inventors, both credited with giving mathematics its greatest push forward since the time of the Greeks.

While the glory of the invention may be great enough for today's scholars to share, it was not enough for Leibniz and Newton, and by the end of the seventeenth century accusations of impropriety were being raised by the backers of both men. The first two decades of the eighteenth century would see the eruption of the calculus wars.

Leibniz had seen some of Newton's early private work, and this was enough to suggest to Newton that Leibniz was a thief. Once he became convinced of this, Newton was largely on the offensive, and he would wield his reputation to great effect. Newton knew that he had invented calculus first, and he could prove it. Still rolling in the glory of his past exploits, he was able to employ henchmen to write attacks against Leibniz, suggesting he stole Newton's ideas, and defending himself against any criticisms that arose. Newton acted not out of empty malice or jealousy but with the firm conviction that Leibniz was a thief. He saw the calculus wars as his opportunity for redemption and a chance for him to reclaim one of the greatest parts of his life's work.

There was no backing down in Leibniz, either. Not one to take such a threat lightly, he fought back, with the aid of followers claiming it was Newton who borrowed Leibniz's ideas. Further, Leibniz worked the community of intellectuals in Europe, by writing letter after letter in support of his own cause. He wrote numerous articles defending himself and multiple anonymous attacks of Newton, and brought the dispute to the highest levels of government, even to the king of England.

At the height of the calculus wars, Newton and Leibniz were attacking each other both openly and in secret, through anonymously written papers and ghost-authored publications. They were both recognized as among Europe's greatest intellects, and both wielded their reputations to maximum effect. Both enlisted trusted colleagues to their cause, and split many of their contemporaries into two camps, championing one or the other. They collected tomes of evidence, wrote volumes of arguments, and were enraged each time they read the accusations of the other. Had Leibniz not died in 1716, the dispute no doubt would have continued even longer, and in a sense the calculus wars did not even stop then, since Newton continued to publish defenses of himself even after Leibniz's death.

Who was right? Newton had a good point in asserting his priority in the invention, and he certainly successfully asserted it. By the time he died, he was recognized not just in England but throughout Europe for having discovered calculus prior to Leibniz.

In England, there still hangs a famous 1702 painting of Newton by Sir Godfrey Kneller in the National Portrait Gallery in London. It shows a middle-aged man in a flowing academic-type brown robe with a slight collar in a brilliant indigo. Newton has large round eyes with slight bags underneath, and the artist brushes pink on his cheeks, nose, and forehead, while blending blue with the flesh tones on Newton's face. The effect is to make him look less ominous than his expression would suggest, though it still seems hard to imagine any humor cracking his serious look.

What could be more truthful—Newton did discover calculus first, ten years before Leibniz did anything. But then again, so what? Leibniz had every right to claim his priority for inventing calculus. He invented calculus independently, and more important, he was the first to publish his ideas, developed calculus more than Newton, had far superior notation (one still used to this day), and worked for years to move calculus forward into a mathematical framework that others could use as well. One could argue as easily that Leibniz's methodology made a greater contribution to the history of mathematics.

Perhaps if Leibniz and Newton had been acquainted under other circumstances, they might have been friends, as they read the same books and both studied the major mathematical and philosophical problems of their day. Leibniz certainly would have loved adding Newton to his vast list of European intellectuals with whom he corresponded regularly over his lifetime. But they never met, and their closest interactions were a brief exchange of letters when they were young men, one letter each during middle age, and another brief exchange when they were old men. Decades passed between their correspondences.

Even though they had few occasions to discourse directly, before the calculus wars began, Newton and Leibniz were given to proclaiming the glory of the other. Perhaps because they had heaped such heavy praise upon each other, their reversal was all the more bitter.

Many writers, including historians and biographers, have dismissed the calculus wars as an unfortunate, even ridiculous, waste of time—perhaps because it reveals the two in their worst light. Leibniz and Newton became downright nasty, and this is hard to reconcile with their otherwise stellar mystiques as ambitious, detached, hardworking, prolific geniuses.

True as that may be, the calculus wars are fascinating because, in them, Newton and Leibniz played out the greatest intellectual property debate of all time—one that, from beginning to end, revealed how these twin mathematical giants, these two elder statesmen of German and British mathematics, were brilliant, proud, at times mad—and in the end completely human.

1

For Once It's Safe to Dream in Color

▪ 1704 ▪

Meticulous, miraculous, ridiculous, fabulous, nebulous, populace, populous, scrupulous, stimulus, tremulous, unscrupulous.

—Word(s) rhyming with "calculus" (pronounced *ka"lkyulus*) with a maximum number of phoneme matches. Taken from www.websters-online-dictionary.org/definition/calculus.

hrẹẹ hundrẹd years ago, history was made when a forgotten English printing press pounded out a few hundred copies of a 348-page work written by a minor government administrator, the retired Cambridge University professor Isaac Newton. Newton was a fairly old man, over sixty, and was already quite famous in England and abroad. But he was not quite the superfamous older scientist he would become in just a few years' time, the venerable elder statesman of British science. In England, Newton's image would approach that of a living god, and, in many ways, *Opticks* helped to create this persona Newton would become.

The book described Newton's experiments and conclusions about the basic physical behavior of light and optics derived through years of independent experimentation. It described such phenomena as how light is bent by lenses and prisms, and how those physical observations lead to a new theory of light and colors: that light was composed of emissions of particles and that white light was a mixture of different rays of distinct colors.

Opticks had a huge impact, and it was well received at home and abroad. It was written in the kind of clear language that only comes from an authoritative and comprehensive understanding of the subject—an understanding that Newton had cultivated over the course of a couple decades. Because it was written in this less formal style, *Opticks* was widely accessible to the reader, and it became a primary text in physics for the next century. The book was subsequently expanded, reprinted, translated into Latin, carried to France and other points on the continent, and sometimes copied out by hand. Albert Einstein once wrote that the world would have to wait for more than a century before the next major theoretical advance in the field after *Opticks*, and the book is still regarded as a classic of physics, still in print, and still read by students of physics today.

A year after the book appeared in 1704, Newton would be knighted by Britain's Queen Anne, and this marked the beginning of the glorious final chapter in his life. He would be celebrated for the rest of his days, admired by intellectuals, kings, and commoners alike. Abroad, he would be a man of celebrity status, recognized by many as one of Europe's premier natural philosophers, a living legend whose company would be sought after by many who traveled to London from elsewhere in Europe and as far away as the American colonies. A young nineteen-year-old Benjamin Franklin tried unsuccessfully to meet Newton in 1725. Forty years later, Franklin had a portrait painted of himself with Newton in the background.

As much of a new beginning as *Opticks* was, it was also the end of an era. Newton was well past his prime as an experimental scientist when it appeared. He was no longer the lonely young genius of half a lifetime before, the silent, sober-thinking lad, as one of his friends described him, who would work day and night, forget to eat, forget to wash, and neglect everything around him except his books, notes, and experiments. He was no longer the man who contemplated the world and figured out how it worked—from gravity and planetary orbits to fluids and tides, revolutionary mathematics, and the nature of light and color. A significant portion of this work was described in Newton's *Principia*, published in 1687, and now, as he was bringing out this second helping, he was much older and busier with professional and social obligations.

In 1704, Newton was no longer a professor at Cambridge and now lived in London, where he would spend the last thirty years of his life as a government administrator in charge of the British mint. His day-to-day business was now overseeing the coining of the English currency, and he threw himself into the mint with all the vigor he had formerly applied to his scientific research. He studied all the parts of the coining process—the machines, the men, and the methods—and became an expert in everything from assaying gold and silver to prosecuting counterfeiters. It was in this role as master of the mint—in a way master of his own universe—that Newton brought forth his book *Opticks* in 1704.

Opticks had been a long time in coming, and publishing it was a catharsis of sorts for Newton. Almost nothing new was published in the book. Much of the material had existed in one form or another among Newton's notes and papers for nearly forty years. Some parts were from lectures he had made as a young professor at Cambridge University and others were taken from letters Newton had written to his acquaintances through the Royal Society in London. Still, before 1704 few people had seen Newton's work on optics.

One of those who had, a mathematician named John Wallis, had tried to get him to publish this material for years, saying that Newton was doing himself and his country a disservice by not publishing it. Wallis wrote to Newton on April 30, 1695, thanking him for a letter and chastising him for not publishing his optical work. "I can by no means admit your excuse for not publishing your treatise of light and colors," Wallis wrote. "You say you dare not yet publish it. And why not yet? Or if not now, when then?"

Ironically, Wallis was dead by the time Newton finally had bound copies of *Opticks* under his arm. Why had Newton waited so long to publish? There were numerous reasons, though none perhaps larger than the bad taste his first attempts at publishing left in his mouth. In the early 1670s, while he was a young professor at Cambridge University, Newton had written a letter on his theory of colors that he sent to be read before the members of the Royal Society in London. His "New Theory about Light and Colors" was published in the *Philosophical Transactions* on February 19, 1672, and it is a letter that reads like one you would expect from the pen of a self-confident young man putting forth a bold new theory to his contemporaries.

For Newton, "New Theory about Light and Colors" was meant to be a third act—a culmination of work already completed. In 1672, he had already been working on his new theories for several years, perfecting his optical outlook on the universe into well-founded science. He had long since gotten over the initial conjectures from which he started, and he was ready to close the book on the work by presenting his conclusions. But Newton was oblivious to the impact that it would have. Writing this letter was something that he would almost immediately regret, because controversy swirled around him after he wrote it. Newton failed to account for the fact that his contemporaries would have to wrestle with the ideas as much as he had for the previous several years. Nor did he suspect how much the people whose theories his was to replace would resist him.

Newton's new way of looking at light threatened the ideas of a number of his contemporaries, including men who were older and more famous than he—for example, his fellow British scientist Robert Hooke. Instead of a third-act curtain call, Newton's letter opened up a whole new dialogue, and he became embroiled in bitter fights with Hooke and others over his new theories—so much so that he swore off publishing for decades. He once even told one of his colleagues that he would rather wait until he died for his works to be published.

Half a lifetime later, after Hooke died in March 1703, Newton was elected president of the Royal Society on November 30, 1703, and it was in this newly appointed role that he published *Opticks*.

The book would be the last original scientific work Newton would ever publish. Nevertheless, it was also a first of sorts because, in it, he staked his claim to the invention of calculus. At the time, most of his contemporaries were attributing calculus to court counselor for the Dukes of Hanover, the German mathematician and philosopher Gottfried Wilhelm Leibniz.

The main body of the book was not about mathematics; it had only a small section in the back on calculus, a treatise Newton had written a dozen years before, entitled *Tractatus de Quadratura Curvarum* (On the Quadrature of Curves). He had written this in 1691, and even then only after the Scottish mathematician James Gregory had sent Newton his own method, which he was about to publish. The essay had started as a letter to Gregory but quickly grew into a text that by 1692 was extensive enough to impress one of Newton's close friends and fellow mathematicians. He revised and shortened this material for publication in *Opticks*. As strange as it may seem for a mathematician as famous as Newton, this appendix was his first actual publication of a purely mathematical treatise.

Newton had discovered calculus during his most creative years of 1665 and 1666. when as a Cambridge University student he had retreated to his family's country estate to escape a particularly bad outbreak of bubonic plague. He had intended to publish his calculus works at the same time as his optical works but, when he published his theory of colors in 1672, he took such a beating from his contemporaries that he swore off publishing in general. Newton was an old man before he published any of his work in calculus, although he wrote letters, sent private, unpublished copies of papers he had written to friends, and wrote page after page in his journals that he never sent to anyone. For most of his life, the heart of his mathematical work was not published.

It might seem strange compared to today's publication-enamored academic world that anyone would sit on an intellectual development as huge as calculus for a period of months, let alone years or decades. Stranger still for someone like Newton, who displayed almost absurd self-confidence at times in his life. And even stranger for a work as important as calculus, which is one of the greatest intellectual legacies of the seventeenth century.

What is calculus? As a body of knowledge, it is a type of mathematical analysis that can be used to study changing quantities—bodies in motion, for instance. Basically, *calculus* is a set of mathematical tools for analyzing these bodies in motion. Given almost any physical motion today (e.g., the movement of clouds, the orbit of GPS satellites around the earth, or the interaction of an HIV drug with its target enzyme), scientists might like to apply the equations of calculus to the bodies in order to predict, track, or model these phenomena.

Differentials are small momentary increments or decreases in changing quantities, and *integrals* are sums of infinitesimal intervals of geometrical curves or shapes. What does that all mean? A nice contemporary way to describe this is to think of the way a baseball curves as it goes from the pitcher's hand to the catcher's mitt. In calculus you express one variable in terms of another. A baseball player throws a perfect fastball, and the radar records the maximum speed, but geometry describes much more—for instance, the changing position of the ball with time. And physics can add another dimension to that, such as accounting for the resistance the ball feels in the air or the effect of gravity on how high the ball is when it crosses the plate or how the spin of the ball will affect the curvature of the pitch. But calculus is about the ability to analyze moving and changing objects mathematically; in other words, using calculus you could calculate all the above without having to throw the ball at all.

Being able to analyze such motion is the domain of calculus. The position, speed, and trajectory of the baseball are changing at every instant as the baseball makes its way to the plate. If you were to take a snapshot of the baseball every hundredth of a second, you could represent the ball's position in terms of time. At time zero, the pitch is on the player's fingertips. A tenth of a second later, it is a few feet in front of the pitcher's hand, another few tenths of a second, the ball reaches its zenith and begins to descend to where it lands in the catcher's glove in the bottom right-hand corner of the strike zone another tenth of a second later—a perfect slider. Newton would have thought of a baseball pitch in terms of these changing quantities as the ball moves.

In the seventeenth century, of course, nobody had heard of or cared anything about baseball. But understanding how the position, speed, and trajectory of a thrown baseball are in a constant state of change is the basis for understanding the physics of all bodies in motion. As such, calculus was the greatest mathematical advance since the time of the Greeks, who had a difficult time getting a handle on such questions. Changing acceleration, for instance, would have been a difficult concept for an ancient Greek mathematician, since it is the measure of the change of velocity over time, and velocity itself is a measure of a change of position with time.

Calculus allowed some of the great problems of geometry to be solved. Newton was not the first to conceptualize such problems. Nor was he the first to successfully tackle the mathematics that could allow him to solve them. The ancients had calculated the area of geometric shapes through what we now call the method of exhaustion—by filling an area with triangles, rectangles, or some other geometrical shapes with easy-to-calculate areas and then adding them up. Using this method, Archimedes determined the area of parabolas and spherical segments.

In the seventeenth century, Johannes Kepler repeated Archimedes' work by thinking of the circle as made up of an infinite number of infinitely small triangles, and then he applied the same reasoning to determine the areas and volumes of other geometric shapes Archimedes never considered. (Interestingly enough, Kepler was inspired in part by the fact that 1612 was a great year for wine but there were not great methods for estimating the volumes of barrels.) Another man, Bonaventura Cavalieri, a friend of Galileo's and professor of mathematics at Bologna, considered a line to be an infinity of points: an area, an infinity of lines; and a solid, an infinity of surfaces.

René Descartes made perhaps the most major contribution to mathematics since the time of the Greeks, when he invented analytical geometry (suffice it to add that the subsequent breakthrough was calculus). Basically, Descartes showed that geometric lines, surfaces, and shapes can be reduced to algebraic equations and that such equations can be graphed geometrically. This was a huge discovery, because it allowed the analysis of geometrical shapes through mathematical equations.

Several mathematicians contemporary to and following Descartes also made contributions. Pierre Fermat, the counselor of the parliament of Toulouse who is most remembered today for his famous last theorem, made a method for finding maxima and minima, drawing tangents to curves so similar to differential calculus that in the eighteenth century some would declared him the inventor of calculus.

Blaise Pascal was a boy wonder in Paris who also worked and wrote on such considerations, publishing his important paper on conics when he was sixteen. Gilles Personne de Roberval worked on geometrical shapes and volumes, and made a general method for drawing tangents to curves. Evangelista Torricelli, a pupil of Galileo, was unaware of Roberval, and published similar results using the infinitesimal method. Scottish mathematician James Gregory in 1668 determined integration of trigonometry functions. John Wallis's book, *Arithmetica Infinitorum*, amplified and extended Cavalieri's work and presented a number of results. Johann Hudde in Holland described a method for finding maxima and minima. Christian Huygens also found ways of determining maxima, minima, and points of inflection of curves. Isaac Barrow published a method of drawing tangents in 1670, and René François de Sluse published one in 1673.

All these works have been called "isolated instances of differentiation and integration," and the mathematicians who accomplished them—along with several more whom I did not mention—were trailblazers. But Newton was the first to figure out a general system that enabled him to analyze these sorts of problems generally—calculus or, as Newton called it, the method of fluxions and fluents. Unfortunately for him, Newton was not the only one to hit upon this.

Leibniz discovered calculus during the prolific time he spent in Paris between 1672 and 1676. Though he was a lawyer and had no formal training in mathematics, he nevertheless showed an incredible propensity toward it. In just a few years he managed to pull together all the mathematical discoveries of his contemporaries to devise calculus. And since Leibniz believed in simple explanations rather than jargon, he invented a completely original and ingenious system of notation to go along with it.

Over the next ten years, he refined his discovery and developed his system of symbols and notations, then published his results in two scholarly papers that appeared in 1684 and 1686. With these two papers, Leibniz could claim intellectual ownership for calculus. He then spent the two decades between those publications and the publication of Newton's *Opticks* refining his ideas, corresponding with his contemporaries, mentoring other mathematicians, reviewing the published work of others, and otherwise extending the techniques of calculus. The word *calculus* was even coined by Leibniz—a calculus being a type of stone that the Romans used for counting.

Calculus was such a promising invention that by the time Newton published "On the Quadrature of Curves" in the back of *Opticks* in 1704, Leibniz was ahead of him by almost two decades. Newton was fighting an uphill battle to wrest credit away from Leibniz, who for over a decade had been basking in the glow of his own invention and was widely recognized throughout Europe as its sole discoverer. Some even thought Newton was plagiarizing Leibniz.

The one place where Leibniz's mathematics had not yet caught on was in England. Part of the problem, apparently, was that the English lacked interest in foreign journals. But this lack of attention in England did nothing to detract from Leibniz's reputation on the continent.

Across the English Channel and in the heartland of Germany, he was at the height of his fame—not only for his mathematical genius but also for his philosophical works.

This short treatise in the back of *Opticks* marked the quiet beginning of the calculus wars because it was the light that revealed the long-hidden feelings of jealousy and resentment between Leibniz and Newton. Newton had suffered in quiet humiliation for years with the knowledge that he was first inventor of calculus, and he was a smoldering fire ready to be released into flames.

On the other hand, "On the Quadrature of Curves" was not the first time someone had made the claim that Newton was calculus's true inventor, but it was the first time that Newton himself had published something to this effect. So Leibniz simply could not ignore it.

—◦◦◦—

IN 1705, AN anonymous review of Newton's essay appeared in a European journal with which Leibniz was closely associated, and it was this review that really fanned the flames. The review made a comment that Newton and his supporters interpreted as a suggestion that the Englishman had borrowed ideas from Leibniz. The German mathematician constantly denied authorship of this review throughout his life but, in the nineteenth century, one of Leibniz's biographers proved that he indeed wrote it. This was not really a revelation, however, because few people ever really doubted that Leibniz wrote the review—least of all Newton.

From the time that Newton read that review and continuing even after Leibniz died in 1716, the Englishman would wage war to stake his claim to the glory of calculus. He would take two approaches. One, quite simply, was to suggest that perhaps Leibniz's own invention was tainted with plagiarism. The other was to assert that in any case he, Newton, had invented calculus first. "Whether Mr. Leibniz invented it after me, or had it from me, is a question of no consequence," Newton would write, "for second inventors have no rights."

Leibniz was not one to take such a threat lightly. He worked the community of intellectuals in Europe by writing letter after letter in support of his own cause. He also wrote multiple anonymous attacks of Newton and published these alongside papers that he wrote, reviewing his own anonymous attacks.

A little more than a decade after *Opticks* appeared, the calculus wars reached their height, and, when Leibniz passed away in 1716, he and Newton were old men fighting openly about which of them deserved credit and whether one had plagiarized the other. Their letters and their private writings are bitter testimonials to their respective brilliance and rival's deceit.

Though it was not until after 1704 that they argued publicly, the foundation of their battle had actually unfolded slowly over the previous quarter of a century, when Newton and Leibniz were much younger. This was an interesting time in history, and the times in which the two had lived played a major role in the dispute that would eventually break out between them. It was a time not just of people coming into conflict but of *ideas* coming into conflict as well. Europe of the second half of the seventeenth century was a world where worldviews were no longer solely the subject of dogma but of debate. Accepted beliefs that had stood for centuries were suddenly felled by the measurements and controlled experiments of the scientific revolution—the birth of the modern in the ashes of the Middle Ages.

In the 1600s, medieval Europe was fading fast, but the continent was still more supernatural than natural. Science and the use of mathematical reasoning to describe the world was emerging in a backdrop that was still seen by most living in those times to be a battlefield inhabited by supernatural spirits—angels and devils that would subject humans to their capricious whims. Dark magic was real. People in the 1600s paid attention to horoscopes, sought omens to predict their fate, interpreted dreams, and believed in miracles. Criminals were detected through divination rather than through investigation. Alchemists tried to transmute lead into gold. Astrologers stood beside astronomers in the palaces of kings. People were accused of witchcraft and hung by their thumbs, whipped, tortured, and treated to grisly deaths. In total perhaps some 100,000 people throughout Europe were accused of witchcraft in the seventeenth century.

The century was also a time that witnessed major political changes, as national identity and nationalism arose alongside the powerful state. In many places, the state became the embodiment of the personal property of the ruler. As Louis XIV famously said, "*L'état, c'est moi*"—*I* am the state. Spoils naturally arose from this point of view; in the 1690s the French regent sold blank patents of nobility for anyone with a bagful of cash. This was, in fact, a common practice throughout Europe in the seventeenth century. Titles and positions were commodities to be bought, sold, and traded as much as they were attainments to be acquired. In fact, King James I of England sold so many knighthoods in the early 1600s that their value decreased—much as you would expect for any commodity that suddenly becomes freely available.

Against this backdrop of occult beliefs, cronyism, and political upheaval, the seventeenth century also saw some of the greatest scientific and mathematical advances made by some of the greatest minds who ever lived. Those hundred years witnessed an explosion of knowledge perhaps unrivaled in the history of civilization. The nature of light and sound were discovered. The diameter of the Earth was estimated to within a few yards, and the speed of light was measured accurately. The orbits of planets and comets were tracked by telescopes and moons were discovered around Saturn and Jupiter. A sophisticated modern view of the solar system evolved, thanks largely to Newton, and it was faithfully described by mathematics. The circulation of blood through the body was carefully charted, and microscopes led to the discovery of cells and a world of tiny organisms too small to be seen with the naked eye.

Because of the wonderment at these achievements, there is a temptation to focus on the intellectual achievements of the seventeenth century. As one historian put it, "During few periods of his history has western man ever really possessed the confidence to believe that by his reasoning alone he could fathom all the questions about himself and his existence."

Nevertheless, we must never forget that calculus and all other significant intellectual developments occurred against a backdrop of horror. If the seventeenth century proved anything about history, it is that it doesn't always unfold gradually.

It was a century of fits and starts; of incredible advances and terrible setbacks; of the most sublime genius and of the cruelest clamoring despotism; and of creative possibility and cruel persecution. For me, the seventeenth century represents a cross between a box of chocolates and a commuter train wreck—an era that delivered to the world a number of remarkable tastes of smooth, sweet, and stimulating hard science and at the same time subjected those living then to the horrors of plague, religious and political persecution, starvation, and war.

2

The Children of the Wars

Be it known to all . . . that for many years past, discords and civil divisions being stirred up in the Roman Empire, which increased to such a degree, that not only all Germany, but also the neighboring kingdoms, and France particularly, have been involved in the disorders of a long and cruel war.

—from *The Treaty of Westphalia*, 1648

Leibniz knew the stink and pain of war from having grown up in a land that was poisoned by it. He was born during one of the most horrible chapters in the history of Europe—the desperate and desolate times during the three-decade-long horror that was the Thirty Years' War. It was a complicated, drawn-out affair involving multiple European states—Danish, Spanish, French, Swedish—that were vying for political gain and German land. The war was long enough that, by the time it ended, it didn't matter much what the causes had been (a complicated fusion of territorial desire and Protestant rebellion). What mattered was that Germany had been utterly shattered by it.

One problem was that the burden of paying for the war was shifted in part from the countries commanding the armies to the lands where the battles were fought. It was no small price. During the Thirty Years' War, assaulting towns and strongholds became difficult, making large, well-organized armies necessary. As a result, European armies swelled to sizes not seen since the times of Julius Caesar— many of the ranks filled with mercenary soldiers. But these large armies meant that there were suddenly tens of thousands that had to be equipped, fed, and perhaps most important, paid.

During the Thirty Year's War, looting was the rule rather than the exception, as poorly paid soldiers would seek their recompense by sacking occupied towns. Moreover, looting became an outright policy for some of the warring armies, implemented so successfully by Sweden's army that, in 1633, the army's expenses cost a fraction of what it had in 1630. And the Swedes were not alone. A Bavarian monk named Mauros Friesenegger quipped, "On 30 September [1633] another troop of one thousand Imperial Spanish cavalry passed through. Although as new recruits they understood no military discipline, they did understand blackmail and robbery."

Nor was the behavior confined to the rank-and-file. For years in the occupied lands, some of the armies' top military and social ranks were occupied by individuals out for personal gain. In Wallenstein's 1632 contract to become general of the Spanish-led army, he held the right to confiscate lands and to grant pardons.

When Leibniz was born in 1646, the war was almost over. He was born in Leipzig, which had been in the heart of the war. In fact, just south of Leipzig was Lutzen, which on the morning of November 16, 1632, about a dozen years before Leibniz was born, had been the site of one of the bloodiest battles of the war. Five thousand men were killed, including Gustavus Adolphus, the king of Sweden, who was cut down leading a blind charge through the fog into the opposing forces.

Two years after Leibniz was born, the war would finally end, when the treaty of Westphalia was signed. The treaty called for a "Christian and universal peace" and a pardon of all war crimes. If peace was Christian, the war had been anything but. Tens of thousands of villages and towns were ruined, and by some estimates, a third of all the houses in Germany were destroyed. Humanity was hit even harder. Perhaps a quarter of the population was killed, and many people were subjected to some of the worst forms of torture and cruelty.

During the Seige of Breisach in 1638, for instance, people were trading furs and diamonds for a kilo of wheat. According to a printed account, "News concerning the Great Famine and emergency that arose during the siege of Breisach," all manner of animals were consumed. Meats that were palatable were sold at an incredible markup and those that were not were still consumed and sometimes traded. "Many mice and rats were sold at high prices," the account reads. "[And] nearly all the dogs and cats eaten." Late into the siege, the residents turned to cannibalism.

Cannibalism is the perfect metaphor for the Thirty Years' War— Europe devouring itself. A man named William Crowne, who was traveling through Germany in 1636, wrote, "From Cologne to Frankfurt all the towns, villages and castles are battered, pillaged and burnt." Industry and commerce did not recover until the eighteenth century, and it has been said that German economic development was thrown back one hundred years.

Leibniz was born at 6:45 a.m. on July 1, 1646, in a home near the University of Leipzig. His parents were Friedrich Leibniz and Catharina Schmuck, both moral and well-educated individuals. Catharina was the daughter of a "celebrated" lawyer in Leipzig, and Friedrich an ethics professor and vice chairman of the faculty of philosophy at the university. Friedrich had been married three times, Catharina his much younger third wife. She was devoted to her two children, Gottfried and his sister.

Legend has it that Leibniz opened his eyes upon the baptismal font, which his father took to be a prestigious sign of the goodness of his being. "I prophetically look upon this occurrence as a sign of faith, and a most sure token," Frederick wrote, "that this my son will

walk through life with eyes upturned to heaven . . . abounding in
wonderful works." Later in life, Leibniz claimed that he had shown
such an aptitude in learning that, even by the age of five, his father
was indulging the "brightest anticipations of my future progress."
Unfortunately, these anticipations were all that Leibniz's father would
ever have. He died in 1652 when his son was only six years old.

One of the things that Friedrich left behind was a library of
books—though Leibniz was not given access to these until an incident with his headmaster at his grammar school. One day Leibniz
found two books that had been misplaced by an older student, and
he began to read them. Leibniz's headmaster was shocked. Though
the books were good texts for an older boy, no boy Leibniz's age
should have been allowed such adult books, the headmaster believed.
He confronted Leibniz's mother, demanding that these books be
taken from Leibniz at once.

Leibniz may have even been flunked into a lower grade had it not
been for a chance benefactor, "a certain erudite and well-traveled
knight," as Leibniz described him. "He disliking the envy of or stupidity of the [headmaster], who, he saw, wished to measure every
stature by his own, began to show, on the contrary, that it was unjust
and intolerable that a budding genius should be repressed by harsh-
ness and ignorance."

As it happened, this nobleman took issue with the headmaster,
arguing that the boy's acute interests in the advanced books was a
sure sign of his budding keen intellect, which should be encouraged
rather than stifled. The nobleman convinced Leibniz's relatives not
only that he should not be punished for reading the inappropriate
book, but that he should be allowed to read all the books in his
father's library at his leisure. "This announcement was a great source
of delight to me, as if I had found a treasure," Leibniz would write
years later in his personal confessions. So, at eight years old, Leibniz
was allowed to enter his father's study. He found books by Cicero,
Pliny, Seneca, Herodotus, Xenophon, Plato, and many others, and
he was free to avail himself of all the Latin classics, metaphysical

discourses, and theological manuscripts shelved there. "These works I seized upon with the greatest avidity," Leibniz said.

Being alone in the study—alone with the books—also awakened in him a love for contemplative independent learning, the sort that he would employ throughout his entire life. He spent many an hour studying the treasures of this library, and he began to read more Latin than a busload of pre-law students at a debate camp. By the time he was twelve, he boasted years later, he "understood the Latin writers tolerably well, began to lisp Greek, and wrote verses with singular success." His Latin was so good, apparently he was able, at the age of eleven, to tackle a difficult assignment in composition in the ancient language in a matter of just a few hours' work. The assignment was to deliver a poetic discourse in the place of his schoolmate, who had fallen ill. "Shutting myself up in my room," Leibniz said, he was able to compose straight through, in a single morning, "three hundred hexameters, of such a character as to gain the [praise] of my instructors."

He was not exposed to mathematics to any significant degree in his early schooling and, as a young man, he would have to teach himself mathematics. He and Newton were similar in this respect.

Their lives paralleled each other's in another way as well. Isaac Newton was also the son of a torn land. In the seventeenth century, England was something of a European oddball in that it was never sucked into the Thirty Years' War—largely because of its geographic isolation from the continent. Britain was also different from many countries in Europe, which were becoming highly centralized states led by a supreme ruler. Instead, it was already highly centralized. If anything, the British monarchy was in danger of losing power rather than consolidating it.

When Newton was born, England's King Charles I had a precarious hold on power. In fact, the country was slipping rather hastily from his grip. The king was warring with Parliament, quite literally, and he resented the check on his power this body represented. He believed in the divine right of kings and thought that he shouldn't be second-guessed or embroiled in petty disputes with Parliamentary officials. During one period of his reign, in fact, he had dissolved parliament for more than a decade, beginning in 1629.

His clash with parliament presented the king with a major financial crisis, though, because the political body had one power the king did not—voting for taxes. He survived for a while by raising fees and fines, but in 1637 a Scottish revolt forced him into the position of needing to raise an army, so he called parliament again.

Five years later, a few months before Newton was born, a civil war between the royal and parliamentary forces erupted. Parliament assumed control of the British fleet, all the major cities including London, and the lands surrounding London. It retained the ability to impose and collect tariffs and otherwise raise funds to supply the war. Charles, on the other hand, financed his armies by pawning off lands, jewels, and other assets. He even took loans from Spain to buy off the Scots.

At the beginning of the war, Charles enjoyed the advantage that his royalist troops were professional soldiers, whereas the parliamentary troops were a rabble. On January 4, 1642, confident in this advantage, Charles stormed parliament: armored, accompanied by armed goons, and intent upon arresting those members of parliament who had earlier defied him. But these opposition leaders were well informed as to the king's movements, and they were gone by the time Charles and his entourage arrived. This was more than just an embarrassing mishap—it was a fatal mistake for Charles and his monarchy. By nightfall, many in the city had gathered together and taken to armed protest, practically making Charles a prisoner in his own castle. Crowds of zealots jeered outside the palace, and the cacophony was impossible to escape anywhere within the palace walls. The situation worsened, and Charles was forced to leave London and escape to more hospitable parts of the country—never to return except for his own execution.

The royalist troops drove up the Great Northern Road, which passed close by the farm where Newton's mother sat pregnant with Isaac. Later the parliamentary troops marched up the same road in pursuit of the king. Though Charles's troops may have been better trained, the parliamentary army led by Oliver Cromwell was disciplined and highly motivated. In the end, Britain's king was executed on January 30, 1649, in London, and his son Charles II fled England a few years later.

Although Newton was born the same year the civil war started, another coincidence is more often noted by his biographers—that Newton was born the same year Galileo died. This has been hailed as significant because in a sense, Galileo was Newton's scientific godfather. Newton would follow in Galileo's footsteps and ultimately describe, using mathematics of his own invention, the physical universe Galileo had observed with his telescope, although an inconvenient fact for anyone who embraces this romantic notion is that Newton was born on January 4, 1643, according to the Gregorian calendar—the year after Galileo's death. England did not follow this calendar in the seventeenth century because Protestants there resisted what they perceived as catholic contamination.

What is perhaps more remarkable than the year of his birth is the way Newton came into the world in the middle of the night so tiny and premature that the women looking after his mother during the delivery thought it a foregone conclusion that he would die—after all, in those days more than a third of all children died before they reached their sixth birthday. Two of these ladies who were sent out to get some medicine for the infant didn't expect Newton to live long enough for them to return. Little could they know that he would outlive them all, not dying until he was more than eighty years old.

Newton's family was unremarkable and largely uneducated. His forefathers were yeoman farmers—not an uncomfortable but certainly a humble lifestyle. His father apparently could neither read nor write, and Isaac was the first Newton who could sign his own name. Newton's father has been described as wild, extravagant, and weak, an interesting guy to know, perhaps, but by the time Newton was born, his father had been dead two months. Newton's father, also named Isaac, died at thirty-seven, just a few months after marriage to Newton's mother, Hannah Ayscough Newton. Hannah, the daughter of a slightly better family, was left as a pregnant widow with a small estate of 46 cows, 234 sheep, and a couple of barns full of corn, hay, malt, and oats near the English town of Westby, in the county of Lincolnshire.

When he was three, on January 27, 1645, Mrs. Newton remarried. Her new husband, Barnabas Smith, was an Oxford-educated clergyman who was the rector of a nearby village. Born in 1582, Smith was sixty-three when the marriage vows were exchanged. The reverend Mr. Smith had his own needs, and the new couple would soon have three children of their own—Newton's half brother and sisters Mary, Benjamin, and Hannah Smith. Newton's mother moved into the good reverend's rectory in North Witham. For whatever reason, Newton did not fit neatly into this picture, so he was sent to be raised by his grandparents in nearby Woolsthorpe, a hamlet in the parish of Colsterworth, in Lincoln County, about six miles south of Grantham. Newton was apparently close to neither of his grandparents, and his attitude toward his stepfather Barnabas Smith was even more volatile. As a child he once threatened to burn his mother and stepfather alive "and the house over them." Newton later regretted saying this, especially after the Reverend Mr. Smith passed away when his stepson was ten, leaving Newton a collection of a few hundred theology books.

At twelve, Newton went to grammar school in nearby Grantham. There, he studied Latin and a few other subjects and boarded in a house that was an apothecary shop owned by a Mr. Clark. It is here, no doubt, that Newton was first exposed to the mixing of chemicals—the spark that ignited his lifelong love for alchemy.

Years later, Newton confessed that he was extremely inattentive in studies and was a bad student. Nevertheless, his vast intelligence, which probably made him seem strange to the other boys, would have been obvious at his grammar school

During this time he was not known to play much with other boys, but rather spent most of his free time by himself tinkering in his room drawing and constructing things. For instance, he was deeply impressed with a nearby windmill being constructed nearby, and he determined to build his own, which he did and was said to be as good as the original. Not satisfied with the fickle blowing of the wind, he built an additional device that allowed a mouse to spin the wheel. He is said to have filled the room with hand-drawn pictures. He constructed a paper lantern that he could fold up and carry in his pocket when he wasn't using it. He also attached the lantern to a kite and flew it at night. He made so many sundials and became so good at it that the neighbors apparently began to come over to see what time it was.

Newton also built doll furniture for a childhood friend—a Miss Storer, who was two or three years younger than Newton and the daughter of Mr. Clark, with whom he boarded. Miss Storer, whose first name has been lost to time, was later to become Mrs. Vincent and is perhaps most famous for describing the young Newton as a "sober, silent, thinking lad." She later confessed to one of Newton's earliest biographers that Newton had been in love with her, though Newton himself left no indication that he had any such feelings.

Building the doll furniture for Miss Storer may have been more interesting to Newton than was Miss Storer herself. This tinkering translated years later into science, and he spent a lifetime building contraptions and conducting experimental work at the same time he did the theoretical work for which he is now so famous.

But Newton had a lot of science and mathematics still to learn—learning that he was not about to do in his grammar school, where he was not exposed at all to any mathematics of consequence. Rather, his grammar school training included Latin and a little Greek. Newton did learn Latin well, which as to be important in his later career, since many of the books of his day were written in Latin.

Leibniz's schooling was equally uninteresting. He later remarked that because his education in mathematics was so poor, his progress was retarded. The scholastic tradition of Germany at the time meant learning Aristotle and logic. Leibniz excelled at logic in school. He claimed that he not only mastered the rules of Aristotelian logic before any other students, but that he also saw some of the limitations of the system.

As a young man Leibniz was to rely upon the method of self-teaching that he cultivated during the years he spent holed up with all his father's books in the old man's library. He was the sort of scholar who threw himself into his work with abandon, gathering his knowledge from books. "I did not fill my head with empty and cumbersome teachings accepted on the authority of the teacher instead of sound arguments," he said once. Another time he reflected that his greatest obligation to his early teachers was "that they interfered as little as possible with my studies."

Leibniz followed in his father's footsteps, studying academic philosophy and law at the University of Leipzig, and he defended his master's thesis, *De Principio Individui* (On the principle of the individual), on February 1664, at the age of seventeen. Leibniz's advisor Jacob Thomasius praised the seventeen-year-old's thesis, declaring publicly that although he was only a teenager, he was capable of investigating anything, however complicated.

Newton was less gifted in practical matters. As a fifteen-year-old, he had to make weekly trips back to Grantham to conduct business. His manservant, necessary because of Newton's young age, was supposed to offer young Isaac advice as he found his way into the world of commerce and adulthood. In fact, Newton was not interested in any such education, and he let the manservant conduct all the business while he busied himself in reading.

In 1659, when Newton was seventeen, he was pulled away from his studies to take over stewardship of his family farm. As the oldest boy, he was expected to become farmer and sheep rancher, and he would have to spend several angry months at home in miserable exile before attending college. But his complete inadequacy for this line of work soon became apparent. Newton's scholarly disposition made him entirely unsuitable to be a farmer of anything but ideas. There is a painting described in a famous biography of Newton that was written in the nineteenth century that captures this time perfectly. It depicts the sheep wandering off, the cattle making feed of the growing crops, and Newton sitting distracted under a tree.

Finally his mother realized that he should be devoted to a life of the mind and sent him back to Grantham for nine months to prepare for university studies. His uncle, the Reverend W. Ayscough, was a Trinity College man and was determined to send Newton there as well. So it came that in June 1661, at the age of eighteen, Newton enrolled in Cambridge University's Trinity College, which historian John Strype called "the famousest college in the university." In these early days Isaac Newton and Gottfried Wilhelm Leibniz knew little of the mathematics that would eventually make them famous, they knew nothing of each other, and they were both inexorably heading toward a similar intellectual destiny.

3

The Trouble with Hooke

See the great Newton, He who first surveyed
The plan by which the universe was made
Saw nature's simple yet stupendous laws
And prov'd the effects tho' not explain'd the cause

—Text from a 1787 engraving entitled
The Most Highly Esteemed Sir Isaac Newton

One of the most succinct, if perhaps overly idolizing, descriptions of Newton's hardworking days at Cambridge appeared three hundred years after he was born. In February 1943, a conference of scholars met in Jerusalem to commemorate the tricentennial of Newton's birth, and an address given by the masters, fellows, and scholars of Trinity College reads in part, "Here [Newton] labored at his calculations and carried out his experiments.... In these precincts he walked in meditation whilst the genius of his mind formed those bursts of experimental activity, when for six weeks at a time the fire in his laboratory scarcely went out night or day."

The image that we have of the young Newton as a superdiligent mad scientist fits because it was true. He had to work so hard as a young man because one doesn't easily uncover the secrets of the universe, which is exactly what he did. As the 1943 masters, fellows, and scholars put it, "Law and order in the physical universe were revealed as never before."

Ironically, the work that Newton did at Cambridge for which he is most remembered is that which he did *away* from Cambridge, during the time he spent holed up in his family home in Grantham. There, he toiled away for many months figuring out how the universe worked and making one spectacular discovery after another while waiting for the school to reopen. It was his so-called *annus mirabilis*, or "miraculous year."

"In those days, I was in the prime of my age for invention and minded mathematics and philosophy more than at any time since," Newton would later write. One of his biographers makes a valid point that the miraculous year should more properly be called his *anni mirabiles*, or "miraculous years," since the time actually encompassed 1665–67.

Often working night and day, Newton rarely placed anything— including food, rest, family, or even his own safety—above his science. He would forget to eat, forget to wash, and grow oblivious to everything around him except his books, notes, and experiments. One story that I like is that Newton's cat grew quite fat munching on all the food that he left untouched. Another is that because he was interested in light and vision, he stared at the sun for periods of time so that he could observe the "fantasies" of color that would be burned into his field of vision. He did this so many times that the story has it that he had to shut himself in a dark room for days to restore his vision.

Worse was Newton's attempts to jab behind his eyeball with a bodkin—a sort of long needle—to alter his retina (the layer of cells with light receptors) and see how this affected his vision. "I took a bodkin & put it betwixt my eye & the bone as near to the backside of my eye as I could: & pressing my eye . . . there appeared severall white, darke & coloured circles," Newton recorded in his notebook, complete with a hand-drawn diagram showing his hand shoving the bodkin behind his anatomically correct illustration of his eyeball.

I saw a copy of this notebook on display at the Huntington Gardens Museum in Pasadena in 2005. While I was standing there, a woman and her teenage son were looking at the book and trying to understand . . . what they were looking at exactly.

"What's a bodkin?" the boy asked his mother.

"It's some kind of needle," she said.

I could see the uncertainty on her face even if her son couldn't, so I jumped in. "It's a long needle that tailors used to use," I said. "It's long, but it has a dull point. They used to use them to poke holes in leather." Nobody said anything after that.

In inventing calculus, which Newton also did during the *anni mirabiles,* he never did anything as severe as almost blinding himself, but in later years he would become blind to the possibility of Leibniz's accomplishments. Newton was a potent mix of brilliance and vanity, and he would later reject the notion that someone else like Leibniz could have accomplished the same thing that he did in these early days.

There is a sense today that the calculus wars were ridiculous because so much of the work that led to the development of calculus and so much of the subsequent work that helped develop calculus into the extensive advanced subject it is today was done by mathematicians other than Newton and Leibniz. Much territory had already been explored in the seventeenth century, and the world was on the brink of finding the calculus. Even though the notion of inevitability of discovery was not as common in the seventeenth century as it is today, when it is quite common for scientists working separately on the same problems to arrive at similar or identical solutions, there is no doubt that calculus was inevitable. All the basic work was done—somebody just needed to take the next step and put it all together. If Newton and Leibniz had not discovered it, someone else would have.

This is not to take anything away from Newton and Leibniz—particularly since they both invented calculus largely by teaching themselves what they needed to know. Cambridge was not a center of mathematics in those days, and Newton was basically on his own. He bought and read a copy of Descartes' *Geometry*. In his later days, Newton recounted to John Conduitt, his nephew-in-law, how he would read Descartes for a few pages, get stuck, go back and reread, get stuck again, read more, and on and on until he had mastered the work.

Newton became familiar with the infinite series. These were ways of finding numerical solutions to problems like the area of a geometrical shape by summing up a series of numbers. In England, mathematician John Wallis had already made progress with this type of analysis by the time Newton arrived on the scene. Wallis is a somewhat obscure figure in the history of mathematics, but he was a mathematical titan of his day, and his work greatly influenced Newton. His book, *Arithmetica infinitorum*, shows some of the first steps in the direction of calculus. In it, he anticipates calculus by seeing the questions that calculus would answer, and he discusses the geometrical ideas of earlier mathematicians who had done some of this work. Reading Wallis's work on infinite series, Newton was inspired to extend this work and invent a general way to analyze geometric curves with algebra—calculus, essentially.

Newton's big breakthrough was to view geometry in motion. He saw quantities as flowing and generated by motion. Rather than thinking of a curve as a simple geometrical shape or construction on paper, Newton began to think of curves in real life—not as static structures like buildings or windmills but as dynamic motions with variable quantities.

By the time Newton was elected a scholar of Trinity College on April 28, 1664, he was aware of the difficult problems that the invention of calculus would solve: those interesting but difficult to deal with problems in geometry, such as finding the area under a curve or finding the tangent (the ability to draw a line perpendicular to any point on the curve). Being a scholar meant that he now had a stipend and living-expensed account, and he was no longer the one to fetch bread. He was, at this point, very close to inventing calculus. Within two years, he would have it. But first an apocalypse would intervene.

A comet appeared in the sky the week before Christmas in 1664, and England's king wondered what it could mean. Charles II, who had been installed a few years before after the failure of the government following Oliver Cromwell's death, was a superstitious man. He followed astrology, was most watchful of such signs, and was more or less representative of his people in this respect. Many in the city wondered what evil fate the comet might portend. William Lilly, a famous astrologer who published a yearly almanac, prophesied in his 1665 edition that another heavenly omen, a lunar eclipse over England in January, would bring "the sword, famine, pestilence, and mortality or plague."

As if that weren't enough, another comet appeared in March 1665. (Actually it was the same comet as the previous December, now on its return trip around the sun.) It's not hard to imagine the fortune-tellers coming out of the woodwork to walk the streets of London in their velvet jackets and black cloaks, bemoaning doom to the people who followed them. And for once, they were right. A horrible plague ravaged England the following summer, and sixty thousand people died in London alone

The fortune-tellers were still quack conjurers and flimflam artists. In fact, it was not a terrible stretch to predict that plague would hit England in 1665, because the plague had already been circulating through Europe for a few years. Holland was particularly afflicted in 1663, when a thousand people a week were dying in Amsterdam; and England was not just geographically close to Holland, it was butting heads with its neighbor across the channel. Britain had fought a recent war with the Dutch and was on the verge of going to war again. Already, in 1662, England had seized the Dutch colony of New Amsterdam and changed its name to New York. Conflict in the colonies could be expected to export conflict back to Europe, and plague to England.

Besides, in those times disease was inevitable. It was as much a part of life of Newton's England as was bad weather. People lived stuffed into slums with poor sanitation. Streets were crowded and had open sewers running down the middle that buzzed with flies in the summer. Half of England's population survived on a subsistence living, and many people suffered from diseases like rickets, caused by a deficiency in vitamin D. People caught measles, malaria, and dysentery in the summer, and in the "r" months, there was typhus, influenza, and tuberculosis, the "captain of all these man of death," as John Bunyan called it. And infections cut across all walks of life. Oliver Cromwell probably died of malaria. Smallpox killed Queen Mary II in 1694. James II may have been stricken with syphilis.

Plague was not necessarily the worst of these diseases because it was not ever-present, as many of the others were. But perhaps because it was episodic it was more terrifying. And to catch the plague is a horrible thing. The infection manifests in painfully swollen lymph glands—called bubos, a term from whence the disease name "bubonic" plague comes. Fevers, chills, exhaustion, headaches, and sometimes severe respiratory illness accompany the disease. Outbreaks during the 1630s killed more than half of the population of some cities. Previous to the outbreak in Holland in the 1660s, there had been an epidemic of plague in France in 1647–1649.

Typically outbreaks of plague occur through the rat population. Large numbers of rats will succumb to an epidemic of infection, and if such a population is living in an urban center, their fleas transmit the bacteria to humans. This is what happened in England in the summer of 1665, when a terrible outbreak of bubonic plague ravaged London. "The contagion now growing all about us," the diarist John Evelyn wrote on August 28, 1665.

By that September, prohibitions against public meetings were in effect everywhere, and by October, one in ten Londoners was dead. "Lord! How empty the streets are and [how] melancholy," Samuel Pepys wrote on October 16, 1665, "so many poor sick people in the streets . . . and so many sad stories overheard as I walk, every body talking of this dead, and that man sick, and so many in this place, and so many in that."

Nor was the plague confined to London. Cambridge, where Newton was in residence, shut its doors in the fall of 1665 because of the epidemic. "[It] pleased almighty God in his just severity to visit this town of Cambridge with the plague of pestilence," as one contemporary account put it. Newton was forced to retreat to the safety of his country home in Grantham, and he stayed there for more than a year until studies at Cambridge resumed in April 1667.

What emerged from these years is arguably the greatest single body of knowledge any scientist has ever produced in such a short time period. Newton arrived at an understanding of the mechanics of motion, and began working on a mathematical description of the laws of motion. He also made major discoveries concerning optics, fluid mechanics, the physics of tides, the laws or motion, and the theory of universal gravitation.

His optical experiments during this time were both beautiful and insightful. He shut himself in a room with no outside light except from a single point source coming from the sun shining through a small hole in the wall. The sun cast a ray of light through the hole, and Newton experimented on the light with a prism. His big breakthrough was understanding that ordinary white light is composed of the spectrum of colors red, orange, yellow, green, blue, indigo, and violet. He also discovered, in careful experiments, that just as a prism can split white light into this spectrum of colors, so can a second prism return the separated colors into white light.

These experiments and others gave Newton the material for his famous book, *Opticks*. But that was not all. Also during this time he conceptualized the material for his more famous book, the *Principia*, which he wrote in the 1880s and which would outline the mathematical underpinnings of physical motion and revolutionize physical science. His law of universal gravitation, described in mathematical detail in the *Principia*, has been called the greatest scientific discovery of all time, and the book continues to be translated from its original Latin today.

It was also in this time that the legend of Isaac Newton and the apple was born. This story is still one of the most enduring tales in the history of science—even though it is probably completely fabricated. Perhaps the only thing that is true about it is that Newton loved apples. The story is no more true than the one about the alligators in the sewers of New York, but it has stuck through the centuries.

Voltaire popularized the tale when he wrote of Newton and the apple almost seventy-five years later. Voltaire's famous story has Newton walking in a garden when he sees an apple fall to the ground from an apple tree branch. This, wrote Voltaire, caused Newton to fall into a profound meditation upon the cause of the apple's falling. According to legend, Newton observed that the apple fell as if it would pass toward the center of the Earth (the center of gravity). Why then, the student-scientist wondered, doesn't the moon fall to earth as well? Perhaps it does. Perhaps it is constantly falling! That, Voltaire claimed, was the inspiration for Newton's theory of universal gravitation, "the Cause of which had so long been sought, but in vain, by all the Philosophers," added Voltaire.

The problem with the apple story is that it oversimplifies the process of discovery that Newton was engaged in. There probably was no one eureka moment (or falling apple moment) that gave Newton the insight the develop his theory of universal gravitation, but rather a less glamorous sequence of long moments spent in study reading, writing, thinking, and working it out. Still, in some ways, it would be easier to understand a genius like Newton if he did simply act as the receiver of great and sudden bursts of insight. It saves having to think about just how he went about the actual work, which strains comprehension. Plus the apple is a great symbol for discovery. Sex, food, sin, and the fall of man—all these things are represented by this humble fruit.

An apple tree planted just to the right of the great gate at the front of Trinity College is said to be a descendant of the actual apple tree that Newton supposedly sat under when he worked out universal gravitation. When I was in Cambridge, I observed more than one person gawking at the tree. Perhaps they were looking, like Newton, for some inspiration to come falling like a red delicious description of nature from those mythic branches. Perhaps, like me, they were puzzled with the disappointment of an apple tree in January—no leaves, no fruit, and scraggly branches that bore nothing but tradition to stir interest. Next to the massive gates of Trinity, it seemed small and insignificant—like it couldn't support the weight of the famous fruit its progenitor produced. Still, apple or no apple, universal gravitation changed science and mathematics forever.

Significantly, Newton also invented calculus in this time—what he called his method of fluxions and fluents. Voltaire's story of calculus is, incidentally, much less interesting than his story of the apple. "'Tis the Art of numbering and measuring exactly a Thing whose Existence cannot be conceived," Voltaire explained simply. Calculus is really a set of rules for analyzing and solving, with algebra, problems related to geometrical curves. It was the answer to some of the big questions of mathematicians of the time—questions like how to find the tangent to (or slope of) a curve at any given point, and how to find quadratures, the areas under given curves.

On Halloween day, 1665, Newton sat down and began to write a short treatise he would call "How to Draw Tangents to Mechanical Lines." A few weeks later he followed this with another paper, "To Find the Velocities of Bodies by the Lines They Describe," which was another early stab at calculus.

I saw a yellowed copy of the manuscript "How to Draw Tangents" under a glass case at the Huntington Library in Pasadena. Most of the crowd walked by with little more than passing glance and seemed more impressed with a calculation of logarithms to fifty-five places— something that Newton worked out in his early days. Newton wrote to an acquaintance about this once, "I am ashamed to tell to how many places I carried these computations, having no other business at that time: for then I took really too much delight in these inventions." The paper has several large triangular columns of numbers—

scary to look at if you are trying to make sense of what they mean, but in the context of a museum, quite striking—even artistic in a visionary sort of way.

Newton wrote a manuscript on November 13, 1665, describing his method of calculus with examples. Over the winter, he continued to work on a number of other topics, and he returned to calculus on May 16, 1666, devising a general method with several propositions for solving problems by motion. Finally, in October 1666, he wrote a tract of forty-eight pages with eight propositions with the heading "To Resolve Problems by Motion, these following Propositions are sufficient." The piece had twelve problems that his methods of analysis could solve directly via his arithmetic methods, including drawing tangents to curves or the instantaneous rate of change (the derivative) at any point along the curve; finding the points of greatest curvature; finding the length of curved lines, finding curved lines whose areas are equal; and finding the area under a curve (the integral) or the area between two curves. This was a real breakthrough.

When he returned to Cambridge in 1667, Newton was a changed man. What he had done, and what Leibniz would repeat a decade later, was to invent one powerful system of mathematics general enough to analyze any curve. At the time Newton was making these discoveries, however, Leibniz still knew nearly nothing of mathematics. On October 2, 1667, Newton received his M.A. from Cambridge and became a Fellow of Trinity College. Strangely, he then set mathematics aside and did nothing more with it for the next two years.

In 1669, he turned once again to mathematics and optics, familiarizing himself with the work of a mathematician in Cambridge named Isaac Barrow. Barrow was the Lucasian professor at Cambridge, a chair founded a few years earlier by Henry Lucas, and Barrow held this chair from 1664 until he stepped down in 1669, passing the distinction to Newton. It had a huge endowment, so Newton got the equivalent of a huge raise and large promotion. Barrow was probably the best colleague Newton could have had, not only for helping him ascend the academic ladder but also because Barrow helped Newton publish, an act toward which Newton had not taken so much as a baby step by the end of the 1660s.

This would all soon change thanks to Barrow and prompted in part by a book published in 1668 by Nicholas Mercator, a German mathematician who lived in London. Mercator's book introduces the term "natural logarithm" and impressively describes how to solve a particular quadrature problem—integration of the function $\frac{1}{(1 + x)}$. This is a trivial problem in calculus today, but it was an elegant and important work when it was published. As impressive as it was, Mercator's work was a specific and rather elementary example of what Newton could solve using calculus. As Voltaire put it decades later, "Mercator published a Demonstration of this Quadrature, much about which Time, Sir Isaac Newton . . . had invented a general Method to perform, on all geometrical Curves."

If Voltaire couldn't help but be impressed three-quarters of a century after the fact, one can only imagine how impressed Newton's contemporaries would have been if they had read Newton's work. But almost none of them could because Newton's work didn't exist anywhere in print. He had written a few manuscripts in the late 1660s and early 1670s that described calculus. The first of these was a Latin work he wrote in 1669, based on his earlier work from 1666, entitled *De Analysi per Aequationes Numero Terminorum Infinitas* (On Analysis by Means of Equations Having an Infinite Number of Terms). This book would later play a crucial role in the calculus wars. Newton and his allies would point to the existence of *De Analysi* as proof that he had developed his calculus years before Leibniz.

De Analysi was supported in second unfinished book that he wrote in the winter of 1670–1671, *Tractatus de Methodis Serierum et Fluxionum* (A Treatise of the Methods of Fluxions and Series). Together, these two books were the first writings that contained Newton's calculus—indeed, the first writings ever to describe calculus. The problem was, he didn't publish them.

Had he published *De Analysi* when he wrote it, Newton would have saved himself a lot of trouble, there never would have been a calculus wars, and he would have advanced knowledge much faster than he did by not publishing. But this sounds easier in retrospect than it was at the time. Publishing such a complicated mathematical treatise would have been extremely difficult in the wake of the great fire of London, which destroyed publishing houses along with much of the rest of the city in 1666—a disaster so dramatic that it's worth describing briefly.

The fire started just after midnight on September 2, 1666, and was apparently the fault of a baker named Thomas Farryner, of old Pudding Lane. But Farryner's fault might have been anyone's. London was a tinderbox of a city in those days. Wooden houses were built upon wooden houses, and their floors were covered in dry straw. The building of new houses within the city walls had continued until the point where every street and open space was filled with a sort of kindling of residential urban decay waiting for a match to march hellfire.

No one could have guessed how devastating the fire would be, though. Surveying the fire on Sunday, the morning after it started, Samuel Pepys called it an "infinite great fire" that threatened to burn the entire city. And a few days later, he lamented, "Lord! What a sad sight it was by moonlight, to see the whole city almost on fire."

John Evelyn bemoaned the dismal sight of the fire in his diary the night after the blaze had started. The next day, he recorded how the fire had worsened: "O the miserable and calamitous spectacle, such as perhaps the whole world has not seen its like since the foundation of it: nor is it to be outdone until the world's universal conflagration . . . God grant mine eyes may never behold the like, who now saw above ten thousand houses all in one flame. The noise, the crackling and thunder of the impetuous flames, the shrieking of women and children, the hurry of people, and the fall of towers, houses and churches, was like a hideous storm. . . . "

"London was, but is no more," Evelyn wrote.

Unfortunately, in the early hours of the fire, the residents of the city were concerned more with saving as many goods as they could than with fighting the flames. The fire could have been contained by tearing down the houses in its path, but this was a tough policy to implement. The lord mayor of London, Thomas Bludworth, refused to tear down buildings without the consent of the owners, and, for obvious reasons, few who owned a house that had not already burned would consent to having their property preemptively destroyed. There were direct ways of fighting the flames—bucket brigades and hand-pumped hoses—but these efforts could do little to quell a conflagration that by Sunday was more than a mile long, blazing a path through the city. Sunday night, and all day and night Monday the fire spread.

By then, it was too late. Panic set in and people began to flee the flames. The streets were sick with carts and conveyances. Londoners of every description—men, women, children, animals—and their property moved toward the city gates and the safety of the outside. The Thames River was congested with barges and boats doing the same. For years, London had been a center pulling in new residents from the largely rural population of England, but now the city was a massive human spout, pouring people back into the countryside.

Pepys, to his credit, succeeded in saving the naval offices and the Tower of London by organizing dockworkers to destroy the buildings around the structures. Other parts of London were saved by using gunpowder to destroy large parts of the city that were lying in the fire's path. But by the time these dramatic measures were taken, it was too late for much of the city. Fueled by strong winds, the flames spread rapidly, and the fate of the city was sealed. By Tuesday, the devastating power of the fire reached the spires of St. Paul's Cathedral, which dominated the London skyline, and burned it to the ground. Rivers of lead melted off St. Paul's Cathedral ran through the streets.

By the time the fire died down, the devastation was massive. Some 373 out of 448 of the city's acres were scorched. An enormous wealth of property was destroyed, along with 13,200 houses and dozens of churches and municipal buildings. About a sixth of the city's population was homeless. And yet, as Voltaire later wrote, "To the astonishment of all Europe, London was rebuilt in three years and arose more beautiful, regular, and commodius than it had ever been before."

The reason why I mention the fire here is not because it's a good cautionary tale in urban planning or because it's an inspirational story about the resilience of a population in snapping back after being beaten down, but because it's a seminal event in the calculus wars. One of the biggest victims of the fire was the publishing industry, seriously damaging the ability of a mathematician such as Newton to publish book-length works. If he were writing a popular pamphlet or clever little handbill, it could have been a different story.

Modern printing was introduced into Europe by Laurens Coster in Holland and Johannes Gutenberg in Germany, and by the seventeenth century, publishing had taken off. The wide availability of books enabled the wealthiest to built libraries, but it also allowed average people to find pamphlets, journals, newspapers, and books on all subjects. Publishing had grown into an industry in Europe, and book sales were exploding there.

Book publishing in London, however, was an industry in crisis when Newton was writing about calculus. Producing a book could be a big risk, since the cost of paper was so high. In the seventeenth century, paper was made from the pulp rendered out of old rags, and the book industry would take big financial hits after plague outbreaks like the one in 1665, because many of the old rags were contaminated by disease and would be burnt instead of pulped, increasing the cost of paper.

Meanwhile, the fire ravaged the city booksellers' stores and destroyed countless stocks of books—so many, in fact, that publishers couldn't afford to take the risk of publishing books that they couldn't quickly sell. As a result, printings rarely exceeded one thousand copies. Typical best sellers of those times were books on religion, for which there was a high demand. This did not bode too well for Newton and other authors of obscure and cryptic mathematics—

especially given all their equations and the difficulty in typesetting them. One book that was published in this time, the optical and geometrical lectures of Newton's mentor Isaac Barrow, is said to have nearly bankrupted the printers.

Thus, for younger and unknown mathematicians like Newton, there was hardly any possibility of publishing a book on mathematics. In fact, *De Analysi* wasn't published until Newton was an old man. Instead, he simply gave a copy of it to Isaac Barrow, and *De Analysi* might have died as a document of no historical importance had it not been for the fact that Barrow was so impressed with it that he wrote to his friend John Collins in London on July 20, 1669: "A friend of mine here that hath a very excellent genius to those things [referring to the book by Mercator], brought me the other day some papers, wherein he hath set down methods of calculating the dimensions of magnitudes like that of Mr. Mercator concerning the hyperbola, but very general."

A few years later, Newton described these methods himself in a letter to Collins he wrote on December 10, 1672, elaborating his approach to finding tangents to curves: "This Sir, is one particular, or rather a Corollary of a General Method which extends itself without any troublesome calculation, not only to the drawing tangents to all curve lines whether geometrick or mechanick or however related to straight lines or to other curve lines but also to the resolving other abstruser kinds of problems about the crookedness, areas, lengths, centers of gravity of curves &c."

Collins was so excited when he read *De Analysi* that he had a copy of it made without Newton's knowledge. This copy would be one of the central documents offered as proof of Leibniz's plagiarism during the climax of the dispute years later.

However hard it may have been for Newton to publish a book in the early 1670s, he still had other options. A new kind of publishing was on the rise—the journal—and in London, the journal *Philosophical Transactions of the Royal Society* had been operating for a few years. It started as a way of keeping track of the papers that were sent to and presented at the Royal Society, and it became a convenient way to publish the latest findings and to keep in touch with discoveries in other parts of the world. This journal was not alone. Several others started in Europe in Newton and Leibniz's lifetimes. In the late 1660s, when Newton was ready to present the world with his work in mathematics, the *Philosophical Transactions* would have been the perfect place to do so. Why didn't Newton have *De Analysi* or some shorter version of it published in the *Transactions*? He may very well have done so had everything gone smoothly for him.

Newton wanted to have his optical works presented first. He would start by revealing to the members of the Royal Society one of his great inventions: a telescope that looked like a toy—an early reflecting telescope. Reflecting telescopes are strange-looking instruments, shorter and fatter than traditional telescopes, with the eyepiece on the side rather than at the back.

The model Newton designed and constructed was less than a foot long, the size of a toy, but size didn't really matter. Barrow demonstrated the reflecting telescope in front of the Royal Society, and it magnified a distant object more than a traditional telescope several times larger. Whereas most small telescopes of the day could magnify objects 12 or 13 times, the much smaller reflecting telescope Newton built could magnify an object "about 38 times," as he wrote in one description. It was a vast scaling down of the technology of the telescope, and it excited members of the society.

"You have been so generous, as to impart to the Philosophers here, your Invention of contracting Telescopes," wrote the secretary of the Royal Society to Newton on January 2, 1672. "It having been considered, and examined here by some of the most eminent in Optical Science and practice, and applauded by them, they think it necessary to use some means to secure this invention from the usurpation of foreigners; And therefore have taken care to represent by a scheme that first specimen, sent hither by you, and to describe all the parts of the Instrument, together with its effect, compared with an ordinary, but much larger, [telescope]."

Newton's reflecting telescope was impressive enough to gain him election to the Royal Society. Thomas Birch, one of the early historians of the Royal Society, wrote in his 1756 *History of the Royal Society of London for Improving Natural Knowledge from its First Rise* that "on December 21, Mr. I. Newton, Professor of Mathematics at the University of Cambridge, was proposed candidate by the Lord Bishop of Salisbury." Newton was ecstatic. On January 11, 1672, an issue of *Transactions of the Royal Society* had a paper that described the design for Newton's reflecting telescope. By that summer, Newton's reflecting telescopes were being built on both sides of the English Channel. Had he done nothing else in his life, Newton would probably still be remembered for this early contribution to optics. But he had so much more to contribute, including his extensive mathematical work, which he could have easily published in the society's journal.

However, he decided he would first follow up his reflecting telescope with a report describing a new theory he had developed on light and colors—something he called "the oddest if not the most considerable detection which hath hitherto been made in the operation of nature."

His theory may have been new, but the field was anything but. Optics had been vibrant throughout the seventeenth century. Descartes had studied optics and so had several figures who followed him, including older and more accomplished scientists than Newton, like Robert Hooke and Robert Boyle in England, and Leibniz's mentor Christian Huygens in France.

Newton's theory was much to the contrary of some of the leading theories of his day, and was a direct challenge to some of these leading scientific minds. To Descartes and others in the seventeenth century, light was like sound—a pulse propagated through a transparent medium, much as sound is really just pressure waves that emanate from a source through the movement of air molecules. Sound ceases to exist in a vacuum, and if you take a bell, stick it in a jar, and pump out the air, it will no longer make a sound when struck. Robert Boyle had demonstrated this to the awe of those who watched, just a few years before. If there is no air, there is no medium to transmit the sound, and many thought that it was the same with light. To Newton's contemporaries, color was not a characteristic of the light but of the vibration in the medium.

Newton was certainly not ignorant of this view and of the body of previous work that supported it. He had read, understood, and had been inspired by the existing theories of light and color. The problem was that once Newton started experimenting, his respect for his own observations outstripped his respect for previous theories. When he saw that the wave theory of light was in conflict with what he observed in his experiments in 1666 and 1667, he boldly proposed that light is not a wave but a particle—an emission made up of innumerable small particles of light traveling through space. He described them as "multitudes of unimaginably small and swift corpuscles springing from shining bodies." Newton also developed a new theory of colors, which held that color was not a characteristic of the wave but a characteristic of the light.

Significantly, he discovered that normal light as we know it is heterogeneous in the sense that it is a mixture of different colors—as we would say today, different wavelengths. White light, Newton found, was far from the pure colorless light that people had always assumed but was rather a combination of all the colors of the rainbow. "The most surprising and wonderful composition was that of Whiteness," Newton wrote in 1672. "There is no one sort of Rays which alone can exhibit this. 'Tis ever compounded, and to its composition are requisite all the aforesaid primary Colours, mixed in a due proportion."

This was exactly the opposite of what many of his contemporaries would have thought. White light to them was the absence of color, just as white paint was the absence of pigment. If you take paints and mix red and green and blue and yellow and violet together, you will get something dark and ugly. So how in the world could white light be a mixture of all these colors as colored light?

It was according to Newton. Replicating his student experiments, he demonstrated this by darkening his room except for a single

source of light, running that point source through a prism and split-
ting it into the rainbow colors, and then running these through a second prism whereby they were recombined into white light. This was
an exciting conclusion—much more so than his mathematical work.

On February 6, 1672, Newton sent a paper describing white light
and his other theories to Henry Oldenburg, the secretary of the
Royal Society in London, to be published in the *Philosophical Transactions of the Royal Society*. Newton's "New Theory about Light and
Colours" was published on February 19, 1672. A copy of the letter
sent by Newton can still be viewed by visitors to the Royal Society
today, as I discovered when I was in London. It contains a cover letter with a florid penmanship announcing, "A discourse of Mr. Isaac
Newton, containing his New Theory about Light and Colors, sent
by him from Cambridge Febr. 6. 1671/72 for ye Secretary of ye R.
Society in order to be communicated to [the body]."

Newton's paper was read to the society on February 8, 1672. The
range of the topics considered by the society on the same day is interesting: after Newton's paper was read, Wallis read a paper speculat-
ing about the moon's influence on atmospheric pressure and on the
barometer. After Wallis, a letter from Naples about tarantula bites was
read, written by an Italian named Cornelio. Next, Flamsteed read a
letter about the moons of Jupiter, and finally a letter from a German
physician, Hanneman, was read, asking about the opinion of the
Royal Society Fellows on sanguification and how it is performed.
Lunar pressure on the atmosphere, toxic spider bites, gas giant moons,
and the ins and outs of bleeding were nothing compared to Newton's letter in terms of the interest generated.

Newton's work was the product of several years of novel and
meticulously performed experiments, analysis, and refinement. He was
not merely describing some part of nature as he saw it, he was seeing that nature be described as it was. His work was an astonishingly
bold new way of thinking about light and colors, and it would eventually be recognized as one of his great accomplishments. Presenting
it was a baby step toward becoming the greatest British intellectual

of his day. In fact, when I was in London, I noticed homage on Newton's tomb in the form of a cherublike creature playing with a prism.

Now a twenty-eight-year-old Cambridge professor, he was ready
to take what should have been a victory lap. But as great an accomplishment as this work would eventually be for Newton, his original 1672 paper instead created trouble. He was forced to endure
stinging public criticisms of his optics work by his contemporaries—
especially Robert Hooke—and Newton did not have the reputation
or prestige that he would later wield against Leibniz to deflect it.

The members of the Royal society showed how seriously they
regarded Newton's work by appointing a committee to look into the
paper and write a report thoroughly evaluating it. Hooke was the one
to write the report, and he included in it his criticisms of Newton's
conclusions. Not coincidentally, the report protected Hooke's own
intellectual territory.

Hooke was the foremost authority in Britain on optics at the time,
and he had been the curator of experiments at the Royal Society for
ten years—a position that he rose to not through politicking but brilliance, especially in his work in optics and the application of optics
to microscopy. Hooke's opinion was so highly regarded in London
society that after the great fire, he was one of a handful of commis-
sioners chosen by the city for the rebuilding effort.

Hooke was also infamous as one of the most outspoken and
intellectually cutthroat of the Royal Society's members and often
wielded the esteem of his position like an ax. In 1672, he set his sights
on Newton's theory of colors, sending the Royal Society a condescending letter claiming to have performed all the experiments
himself, prior to Newton. In addition, he concluded the experiments
proved that light was a propagating pulse through a transparent
medium and color was a refraction of light—exactly what Newton's
work was supposed to be refuting. In other words, Hooke claimed
that the difference was not of data but of the interpretation of data.

"I have perused the Excellent Discourse of Mr. Newton about colors and Refractions, and I was not a little pleased with the niceness and

curiosity of his observations," Hooke wrote, "[But the experiments] do seem to me to prove that light is nothing but a pulse or motion propagated through an homogeneous, uniform and transparent medium, and that color is nothing but the disturbance of that light by the communication of that pulse to other transparent mediums."

This letter, which Hooke had taken all of three or four hours to write, must have been a smashing blow to Newton. Hooke was one of his heroes, and Newton had been greatly influenced by Hooke's famous book, *Micrographia*, his seminal studies of the microscopic world—a book that Pepys called "the most ingenious book that I have ever read in my life." When Newton had read *Micrographia,* he had been fascinated by the detailed drawings of lenses and lengthy discussions of optics inside the book and had recorded pages and pages of notes on it.

After reading Hooke's 1672 letter condemning him, Newton spent three months composing a reply, carefully going over his notebooks and other materials and pulling together many different lines of thought to address Hooke's criticisms in a single lengthy discourse. The brash young twenty-something scientist confronted his elder head on. He wrote pages and pages addressing Hooke's criticisms point by point. After a few months' delay, he sent a highly edited version. As in so many other times in his life, Newton showed that his best defense was a strong offense. He opined that Hooke's theory was "not only insufficient, but in some respects unintelligible."

Newton essentially believed that objections without experimental results should be rendered invalid. And he had done the experiments. Once separated into component colors, the various colors of light could not be further separated or changed by passing them through a prism.

"I have intercepted [a single colored ray of light] with the colored film of air interceding two compressed plates of glass; transmitted it through colored mediums, and through mediums irradiated with other sort of rays, and diversely terminated it, and yet could never produce any new color out of it," Newton wrote in his paper. "It would by contracting or dilating become more brisk, or faint, and by the loss of many rays, in some cases very obscure and dark; but I could never see it changed in specie."

Newton was not the only one who had a hard time getting his novel theories accepted—based on experiments though they were. In fact, this was a common theme in the seventeenth century. Johannes Kepler's theory that the planets follow elliptical orbits was a hard pill for many of his contemporaries to swallow. Circles were more perfect shapes, the criticism went, so what need would the heavens have of ellipses? This same kind of thinking caused many to question the existence of sunspots after Galileo discovered them. Why would the sun have spots? Galileo faced a similar protest of his discovery of moons that circled Jupiter. Because these moons were invisible to the naked eye, Galileo was ridiculed by at least one Italian scholar, who said in effect that if we couldn't see them, they would be of no use to us, and therefore couldn't exist. The critic also made a complicated argument that involved the number seven. New moons would increase the number of planets and moons in the solar system above seven. But there could only be seven planets for the sake of natural harmony—just as there were seven orifices on the human head.

Not all resistance to new ideas was so banal. These were dangerous times for ideas as well as their authors. The inquisition in Rome placed Galileo under house arrest for life, and, after publication of his *Dialogue* in 1623, banned him from ever publishing again. Descartes left his native France in 1628 for fear that he would be persecuted for writing unpopular ideas, and he stayed in self-imposed exile in the Netherlands until 1644. John Bunyan, who wrote *Pilgrim's Progress*, the so-called layman's bible that was one of the most famous books of the seventeenth century, was locked up from 1660 to 1672 for the seemingly innocuous charge of preaching without a license. Giordano Bruno was burned at the stake in 1600 for daring to put forth unpopular positions.

Newton never faced anything as harsh as burning at the stake, but there is no question that Hooke's attacks clouded his psyche for decades. And Hooke was not alone in opposing him.

In the months after Newton sent in his paper on colors, other criticisms drifted in from the continent, and Newton responded with a number of letters. He got comments from a Jesuit priest, Father Ignatius Pardies, who was a respected member of the Paris community of scientists. Pardies protested that he simply could not believe that colored rays combined should make white light. His comments were intelligent, valid criticisms that Newton was able to address in kind. Likewise, intelligent comments came in from Huygens, Leibniz's Paris mentor. However, criticisms of a different nature came from a Belgian named Franciscus Linus, whose greatest legacy seems to be being remembered as a stupid, ignorant, and narrow-minded man.

The effect of the criticism, comments, and correspondence on Newton was to send him back into his turtle shell of Cambridge. He even intimated to Oldenburg that he would prefer to drop out of the Royal Society, and was considering abandoning all experimental research.

The unfortunate victim of all this fighting was Newton's work on calculus, since Newton always intended to publish his optical and calculus work together. The pain of publishing the former caused him to abandon plans to publish the latter. Because of the trouble with Hooke, Newton lost his taste for publishing altogether. If there was any possibility of his publishing his mathematical works before, there was no longer any question that this could not be done. Though he had invented his fluxional calculus in the mid-1660s, the world would have to wait another two decades before it got a taste of it. And when it did, Newton would not be the author. Until then, Newton became a sort of Greta Garbo of the science world.

Events were transpiring in Europe—a war for France and much of the rest of the Continent was looming on the horizon—that would steer Leibniz first to Paris and then to London—and into a collision course with Newton. Leibniz would display none of Newton's reservations about publishing or sharing his ideas with others.

4

The Affair of the Eyebrow

Many of his dreams have been realized and have been shown to be more than the fantastic imaginings that they seemed to all his successors until the present day. . . .

—Bertrand Russell in the 1937 preface to his
critical exposition on Leibniz's philosophy

For most of his life, Leibniz rarely worried about being over-shadowed by Newton or anybody else. He was one of the most prolific thinkers of his day, and his far-flung interests led him to contribute advances in fields as diverse as medicine, philosophy, geology, law, physics, and of course mathematics. It was exactly the sort of ambition that led Leibniz to plunge into mathematics in the early 1670s—not simply to understand everything that had been done by his contemporaries, but also to synthesize everything known at that time into one general system that could serve as a tool for future discoveries.

Mathematics was not his main interest in his early days. In fact, Leibniz did not plunge far enough into mathematics to invent calculus until he was nearly thirty. And even then, calculus seemed but one facet of his grand vision for knowledge in general. He saw all human ideas, concepts, reasonings, and discoveries to be a combination of a small number of simple, basic fundamental elements—like numbers, letters, sounds, colors, and so on. Leibniz hit upon the idea of creating a universal system that would provide a way of representing ideas and the relationships among them—an alphabet of human thought with which ideas, no matter how complicated, could be represented and analyzed by breaking them down into their component pieces, like the letters that make up w-o-r-d-s a-n-d s-e-n-t-e-n-c-e-s.

The *characteristica universalis* or alphabet of human thought was first attempted in his doctoral thesis, *Dissertatio de Arte Combinatori* (Dissertation on the combinatorial art). A little later in life, he described his idea in the most visionary and optimistic terms: "Once the characteristic numbers for most concepts have been set up, the human race will have a new kind of instrument which will increase the power of the mind much more than optical lenses strengthen the eyes and which will be as far superior to microscopes or telescopes as reason is superior to sight. The magnetic needle has brought no more help to sailors than this lodestar will bring to those who navigate the sea of experiments."

Such a reduction of complex ideas may sound foolishly simple, but the attempt to come up with an alphabet of human thought is what led Leibniz to calculus. He knew little mathematics when he wrote his "Dissertation on the Combinatorial Art," but in a way the dissertation prepared him to discover calculus because it allowed him to appreciate the need that calculus would fulfill. Calculus, after all, is a body of knowledge dealing with the analysis of geometry and numbers, and for Leibniz this was one example of a larger logical system for analyzing all his *characteristica universalis*.

Moreover, "Dissertation on the Combinatorial Art" had a very direct impact on the calculus wars because it set into motion a sequence of events that would lead Leibniz to Paris, where he would invent calculus, and to London.

However brilliant his work, he was denied his doctorate at the University of Leipzig in 1666. Why this occurred is not entirely clear. One of the stories is that the wife of the university's dean convinced her husband not to award the doctorate to young Leibniz for some personal reason. But perhaps he simply fell victim to the academic politicking of the university. There were a limited number of spots available for graduation and, had Leibniz's thesis been accepted, he would have prevented a more senior student from graduating.

Undeterred by this setback, Leibniz left Leipzig, matriculated to the nearby University of Altdorf in October 1666, and graduated from there a few months later, receiving his doctorate from the university in February 1667. His dissertation, *De Casibus Perplexis* (On difficult cases [in law]), held that the law had to answer a certain number of uncertain cases, which in his day were often decided by drawing lots and other arbitrary means. Leibniz argued that these difficult cases should instead be decided by reason and the principles of natural justice and international law.

He claims that his thesis dazzled the audience. "I received the degree of a doctor from the University of Altdorf, with great applause," Leibniz once bragged. "In my public disputation, I expressed my thoughts so clearly and felicitously, that not only were the hearers astonished at this extraordinary and, especially in a jurist, unexpected degree of acuteness; but even my opponents publicly declared that they were extremely well satisfied."

Following the awarding of his doctorate, the education minister at the university, a man by the name of Johann Michael Dilherr, told him that he could guarantee Leibniz a professorship if he was so inclined. Leibniz was not. "My thoughts were turned in an entirely different direction," he said later in life. "I gave up all other pursuits and confined my attention exclusively to that occupation upon which I was to depend for a livelihood."

What was this livelihood that caused Leibniz to reject the offer? It was an occupation through which he sought to do something more practical—work that would confer the greatest benefit to mankind.

He decided to pursue law. The thought that a lawyer would have more opportunities to do good than would a university professor would no doubt make many modern university faculty members laugh or wince. Nevertheless, once Leibniz finished his doctorate in 1667, he left university life forever. He would face the world, an ambitious and brilliant young lawyer with a keen interest in politics and learning, but not much knowledge of mathematics.

He settled in nearby Nuremberg, and had no trouble fitting into the learned societies there. One of the groups he became acquainted with was an alchemical society. The story is that he wanted to gain access to their society and secrets but, since he was an outsider, he did not have a way in. So he devised a plan. He consulted the most difficult alchemical textbooks he could find and wrote down the most obscure words that they contained, and he came up with a paper that was both impressive and meaningless. He later admitted that it made no sense whatsoever, even to him. But he so impressed the alchemists at his ability to write profoundly that they gladly welcomed him into their society and made him their secretary. For months, he joined them in discussion and debate. Later, though, he was to denounce the cult of alchemy as the "gold-making fraternity."

In 1667, Leibniz's life took a dramatic turn. He met a wealthy and well-connected German statesman, Baron Johann Christian von Boineburg, a man of prestige and learning known in many of the German capitals. In the next five years, Leibniz became a close friend of Boineburg, serving as his secretary, assistant, advisor, librarian, and lawyer for several years. This relationship would prove crucial in Leibniz's life because it would be Boineburg who would convince him to go to Paris a few years later.

The baron saw in Leibniz a great protégé, and from the beginning, his assistant's intellect impressed him. Boineburg once wrote to an acquaintance introducing Leibniz in the grandest of terms. "He is a young man from Leipzig, of four and twenty," Boineburg wrote. "Doctor of laws and learned beyond all credence."

Boineburg helped Leibniz get into the good graces of the archbishop elector of Mainz, Johann Philipp von Schönborn, who was a regional political leader of some prominence. During this time, Germany was something of an amalgam of states, dozens of which were ruled by bishops and archbishops like Schönborn. Mainz was a German state but also like a small country, in that it was part of the Holy Roman Empire. (Voltaire once quipped that the Holy Roman Empire was neither holy nor Roman nor even an empire.) Boineburg had been close to the archbishop and was formerly a minister of the court at Mainz (he was fired in 1664, but shortly thereafter reconciled with the elector after his daughter married Schönborn's nephew).

This meant that Boineburg was well positioned to introduce Leibniz to Schönborn. Leibniz wrote an impressive essay, "A New Method of Teaching and Learning Law," which is said to be rich in ideas. Boineburg convinced him to dedicate it to Schönborn and arranged for Leibniz to have an audience with the archbishop, to present his essay to him in person. Schönborn's response was to make Leibniz a judge in the High Court of Appeal at the age of twenty-four.

Leibniz was assigned to work with a man named Herman Andrew Lasser on a project revising the legal code. Together they wrote a large work, Leibniz writing two parts and Lasser also contributing two. Leibniz's opened powerfully: "It is obvious that the happiness of mankind consists in two things—to have the power, as far as is permitted, to do what it wills and to know what, from the nature of things, ought to be willed." Some modern, some antiquiated, Leibniz sought to find a systematic basis for this diverse set of laws.

Legal reform was a hot topic in those days because the Holy Roman Empire was complicated by an intricate system of laws that varied from state to state. One effect this had was to fractionate Germany, and because the various states acted autonomously, various rulers considered only themselves when deciding with whom to form alliances. Since Germany was centrally located in the middle of Europe with bordering states on the east, west, north, and south, these alliances were key.

Moreover, a number of uncomfortable divisions had arisen out of the reformation following Martin Luther's introduction of Protestantism more than a century before. States were divided between the Protestants and the Catholics. The Peace of Augsburg in 1555 allowed local princes to determine the religion of the land, but it only further divided Germany and subjected states to the whims of their rulers. Perhaps the most dramatic example of this was in German in the state of Rhineland-Palatinate, which switched from Catholic to Lutheran in 1544, from Lutheran to Calvinist in 1559, from Calvinist back to Lutheran in 1576, and from Lutheran again to Calvinist in 1583.

During the five years that Leibniz was a close advisor to Boineburg, he had his first taste of ambassadorial politics. When John Casimir, the king of Poland, stepped down from the throne in 1668, a number of people aspired to take his place. One of these, the prince of Neuberg, was supported by Boineburg, and Boineburg asked Leibniz to help toward this end. What Leibniz did in response was to write a pamphlet in which he not only gave the merits of Neuburg's cause but also investigated the nature of Poland in general—its government, its conditions, and so forth. Although Neuburg did not become king, Boineburg rewarded Leibniz by recommending him to be a member of the elector of Mainz's council.

It was through his relationship with Boineburg that Leibniz was thrust into Paris, London, and eventual conflict with Newton. By the beginning of 1672, the drums of war were deafening as France, Europe's main superpower, was once again turning an aggressive eye toward other European countries. Louis XIV was furious with the Dutch, who had been his allies, because in 1668, Holland joined with England to thwart France's attempt to annex the Spanish Netherlands. This set off a commercial dispute, with France slapping heavy tariffs on Dutch goods. By 1671, the situation was dire and Europe was on the brink of what could be another major war.

This created a confusing scenario where many of the states in Germany had various alliances with or against France. Johann Friedrich, the Duke of Hanover, was a good example of this. His foreign policy was to support France in exchange for money. But all alliances would be truly tested after France began amassing troops along its eastern borders as it prepared to invade Holland.

Schönborn was forced to abandon his alliance with the Duke of Lorraine after the duke pushed the elector to form an alliance with England, Holland, and Sweden against France at a meeting that took place in July 1670. Boineburg and Leibniz were both at this meeting, and they both opposed the prospect of such an alliance.

Leibniz even wrote a pamphlet with the unwieldy title, *Reflections upon the Manner in which, under Existing Circumstances, the Public Safety, Both Internal and External, May Be Preserved, and the Present State of the Empire Be Firmly Maintained.* This pamphlet warned of the dangers of taking sides against France, and Schönborn heeded the advice, sitting idly by as tens of thousands of French troops poured into Lorraine, and the Duke, his erstwhile ally, was forced to flee.

Boineburg saw the stupidity of standing against the great military superpower of France. Besides that, he had to much to gain from keeping Mainz on France's good side—he had property and a pension in France that were owed to him, and he believed he could recover this small fortune if he played his cards right. He hoped to be sent to France to collect his money and, at the end of 1671, just as Newton was preparing to present his new theory of light and colors, Boineburg was positioning himself to go there.

But things became unglued when the French foreign minister died. It was several months before a new minister, Simon Arnauld de Pomponne, was appointed in January 1672. By then, a French ambassador had arrived in Mainz on a mission to ask for free passage of his war ships on the Rhine River so that Louis XIV's troops could attack Holland more easily. The presence of the French ambassador in Mainz made Boineburg's trip to Paris irrelevant. So Boineburg decided he would send Leibniz to Paris instead.

Leibniz drafted a rather vague document and sent it to Louis XIV on January 20, 1672, mentioning how he and France could benefit from "a certain undertaking" that had advantages for France. He gave no details as to what this undertaking would be, and the document must have piqued a great deal of curiosity in France and in the new French foreign minister, Simon Arnauld, Marquis de Pomponne, because a reply arrived on February 12, 1672, asking Boineburg to come and present his proposal. Boineburg sent word on March 4, 1672, that he would send Leibniz in his place.

Leibniz's plan was bold almost to the point of being far-fetched. He wanted to convince Louis XIV that France should not go to war with Holland, by making a case for how profitable it would be to instead turn his country's aggression and ambitions toward Ottoman Empire–controlled Egypt. Egypt, with its command of important trading route points, was a much more lucrative target, Leibniz argued, and attacking the Ottomans in Egypt would also shore up the eastern portion of Europe, where such cities as Vienna were under threat of attack from the east.

It may seem strange to propose an invasion of Egypt as a plan for peace, but the idea of turning war within Europe toward the outside world was nothing new. In the 1300s an Italian, Marino Sanuto, wrote a book, *Secreta Fidelium Crucis*, which proposed essentially the same thing to the pope. In fact, Leibniz drew upon this centuries-old work when he came up with his up-to-date version of the plan. But in the initial letter, none of the specifics were presented. In fact, it was so lacking in detail that nowhere did it even mention the word "Egypt."

Leibniz and one servant set out for Paris on March 19, 1672, to present his eleventh-hour appeal. He carried with him a letter of attorney from Boineburg, a letter of introduction, traveling expenses, and a sincere desire to convince the French king of the value of forgetting about war in Europe and instead turning to parts of the non-Christian Middle East. His mission was kept semisecret, and he traveled under the cover of representing the personal interests of Boineburg, arriving in Paris at the end of the month.

Leibniz must have been excited about the trip, like any young man on his way to the big city for the first time. Paris was one of the largest and grandest cities in Europe, and it was the playground of Europe's rich and elite. Even though much of Germany was at war with France

at some point or other throughout his lifetime, France was nevertheless a model of seventeenth-century courtly life. Its features were to
be admired and its courtly manners to be imitated in as many lavish
details as possible.

Moreover, Leibniz was going there to present a proposal to the
highest levels of French government. This was very appealing to him
because one of the things he liked to do was act ambassadorial. He
might have entertained ambitions of actually being an ambassador,
but he lacked the one crucial asset that would have allowed him to
do so—the pedigree of a high birth. He may have been representing Boineburg, but he was himself no Boineburg. Nevertheless,
there was a real possibility that he would be presenting his work to
Louis XIV, who was a fabulously powerful monarch.

Louis XIV had been a boy king, inheriting the throne from his
father when he was only four years old. Because he was a child, he
was completely unprepared to rule, and a regency government was
installed instead, with Louis's mother as regent and her close advisor, the Cardinal Mazarin, in charge for the next dozen years. When
Cardinal Mazarin died, Louis XIV took over and became the longest-
ruling king in the history of France. He was the model of the
absolute monarch: Though he governed France with the help of
myriad advisors and confidants, he kept absolute power, and if any
one person had the power to change the course of history at will and
to stop a war upon hearing a petition, that was Louis XIV.

As a military strategist, Leibniz was more than a century ahead of
his time. France would indeed eventually invade Egypt under
Napoleon, who grasped the value of the peninsula exactly as Leibniz had suggested. In fact, when Napoleon invaded German and
occupied Hanover in 1803, he was annoyed to learn that Leibniz had
anticipated him by more than a century.

However flattering this might have been for Leibniz had he have
known, the proposal was an ill-timed flop in his lifetime—as it
turned out, he never had an opportunity to present his proposal.

On April 6, 1672, Louis XIV and his subordinates published a

short document, "Declaration of War against the Dutch." Issuing it
from Versailles, the king ordered it disseminated throughout France
and its dominions where, as all his subjects were to read, he commanded them to "fall upon Hollanders." With the French already in
position to invade, the Dutch were forced to open up dikes and flood
the countryside to slow the French advance. The Franco-Dutch War,
as it is called, had begun, and it would drag on for the next six years.

When Leibniz finally did arrive in Paris, his original proposal was
now moot. Nevertheless, Leibniz and Boineburg, keeping in communication, did not abandon their plan, but modified it making invasion of Egypt an enticement to end the war as opposed to a proposal
to prevent it. Once the battles within Europe were concluded, they
proposed, an invasion of Egypt could begin. Leibniz wrote a paper
to this effect, "Consilium Aegyptiacum," which is said to argue the
case with eloquence, learning, and mastery.

To bolster their case, Leibniz and Boineburg brought the elector
of Mainz on board. Schönborn thought it was a great idea, and he
immediately sent word to Louis XIV, who was encamped with his
army at the time, offering to mediate peace so that the French could
quickly set sail for North Africa. The answer was, in effect, an eloquent "no thank you, the crusades are over." "As to the project of the
holy war, I have nothing to say," read the response to the German
court. "You know that since the days of Louis the Pious, such expe-
ditions have gone out of fashion."

But for Leibniz, the crusades were just beginning. He decided to
make the most of his time in the French capital anyway. What an
opportunity this was for him! In Paris, he was alone and without
major day-to-day duties. After spending several months learning
French and setting himself up in this new urban setting, he buried
himself in the libraries for days; also, because he arrived in Paris as
the representative of Boineburg and carried with him letters of
introduction, many doors were opened to him.

With these open doors came numerous opportunities, and, in the
few years he spent in Paris, he was able to partially support himself

with legal work. He was, after all, a lawyer who could bring skills to the elite society, drawing up documents, taking legal briefs, or providing other services and actions on behalf of the well-to-do. He secured the release of a foreign prince from jail, for instance, and he arranged for the divorce of the archduke of Mecklenburg—a man who had been hated by his subjects at home, which forced him to flee Mecklenburg in 1674 to Paris, to which his temperament was more suited. However, Mecklenburg had converted to Catholicism, which presented him with a problem. Before, when he was Protestant, he had no trouble divorcing his first wife. But now that he was married to a nice Catholic lady, divorce was not so easy. So Leibniz helped him out.

Leibniz was busy enough with this type of work and other social obligations. In what would become a theme throughout his life, he became distracted from what he saw as his more interesting intellectual work. "My mind is burdened by a great variety of labors, in part required of me by my friends, and in part by persons of rank," he wrote to the secretary of the Royal Society in the summer of 1674. "Therefore I have much less time than I could wish to devote to the study if nature and to mathematical investigations. Nevertheless, I steal as much of it as I can. . . . "

Luckily, Paris was an intellectual capital of Europe and boasted some of the finest minds alive in Leibniz's day. He met many of them there, and was inspired to come up with highly original, although sometimes impractical, ideas—such as a way to determine longitude; a pneumatic gun; a concept for how a boat might be able to dive, submarine-style, to escape from pirates; and an idea for improving watches. In Paris, Leibniz truly began his lifelong career of scholarship, acquiring a breadth of learning and acquaintance that covered the whole of the "republic of letters" as the philosopher Bertand Russell once described it. And he made a few discoveries in mathematics.

His journey on the road to discovering calculus began in the fall of 1672, when he met Christian Huygens. A Dutch physicist and mathematician, Huygens was the son of a famous literary and diplomatic figure in the Netherlands, and he had something of a gift for words himself, once declaring, "The world is my county," and adding that promoting science was his religion. His father was a friend of Descartes, and Huygens was a strict Cartesian for his entire life, which had a strange influence on his work at times. For instance, after he discovered a moon of Saturn, he stopped looking for more moons in the sky because Cartesian symmetry held that, since these were six planets, there should be six moons.

Despite how silly that reasoning seems today, Huygens is still regarded as one of the greatest scientists in the seventeenth century. When Leibniz first went to visit him, Huygens was perhaps the foremost natural philosopher living in Paris and one of the best-connected intellectuals in Europe. As a measure of how great a mathematician and scientist Huygens was, even though he was Dutch and living under the highly xenophobic regime of France's Louis XIV, Huygens was still the leading member of the Académie des Sciences, an organization that he had helped found.

Huygens's status was well deserved. A gifted craftsman who developed methods for making lenses in the mid-seventeenth century, he made several important contributions to science in his lifetime. In 1655, using his improved lenses in his telescope, he observed the rings of Saturn. Master of the latest mathematics, Huygens studied the pendulum, analyzed it mathematically, and used it as an engine to drive clocks of his own invention.

Huygens and Leibniz hit it off right away, and in the next few years they became friends. More importantly, the older, wiser Huygens became Leibniz's inspirational mentor, encouraging the German to look deeply into mathematics. "I began to find great pleasures in geometrical investigations," Leibniz wrote years later, as he remembered that time in a letter to the Countess Kielmansegge near the end of his life.

Huygens must have gotten a great deal of pleasure out of this interaction as well, because his protégé was beginning to make rapid progress by the end of 1672. That fall, Huygens gave Leibniz a challenging problem involving the sum of a mathematical series, specifically the sum of an infinite number of fractions, each smaller than the last: $1 + \frac{1}{3} + \frac{1}{6} + \frac{1}{10} + \frac{1}{15}$, and on and on. Huygens asked Leibniz to calculate the sum of the infinite series; Leibniz sat down and was able to come up with a solution (the answer is 2). Huygens was impressed and urged Leibniz onto further studies, suggesting books that the younger man should study. One of these was *Arithmetica Infinitorum*, by the English mathematician John Wallis, which had so inspired Newton just a few years before.

Another book was by the Belgian Jesuit mathematician Gregory St. Vincent, which Leibniz borrowed from the Royal Library in Paris and began to study as soon as Huygens suggested he read it. St. Vincent thought of a geometrical area as being the sum of an infinite number of infinitely thin rectangles. This work anticipated integral calculus, the second side of the calculus coin that can be used to determine the area or volume of a geometrical shape by applying a set of algebraic tricks that essentially add up all these little triangles.

Leibniz also read Bonaventura Cavalieri, a friend of Galileo's and professor of mathematics at Bologna. Cavalieri had developed the idea of the indivisible—a small section of a geometrical shape which, when taken with all the other small sections, would constitute the initial shape itself. He considered a line as being made up of an infinity of points, an area an infinity of lines, and a solid an infinity of surfaces. Think of this as a stack of pancakes: The stack is made up of all the individual flat pancakes. Cavalieri's 1635 book, *Geometria*, proved such facts as that the volume of a cone is one-third the volume of the cylinder that fits around it.

While studying these works, Leibniz started to go further and do some original mathematics, which he thought of publishing in a French journal until it unexpectedly folded. Aside from this minor setback, by the end of 1672 Leibniz was beginning the most outstandingly productive times in his life—certainly the greatest time he spent considering mathematics. In his four and a half years in Paris, he grew from a lawyer with little formal training in mathematics into a scholar who not only understood the furthest mathematical advances of his contemporaries but pushed them forward—for example, inventing calculus.

However, during this time, Leibniz would also feel the biting sting of personal defeats, the first of which came less than a year after he arrived in Paris, when Boineburg died on December 15. This was not just the loss of a patron. Boineburg, whom he later called one of the greatest men of the century, was someone for whom Leibniz had great respect and affection.. And this was not the only death that Leibniz would have to deal with. A month after Boineburg's demise, Leibniz's sister died.

But perhaps the greatest personal defeat would come a few months later on a trip to London in the winter of 1673, where he headed on another diplomatic mission in early 1673 with Boineburg's son-in-law, Melchior Friedrich von Schönborn, the nephew of the elector of Mainz. Young Schönborn showed up in Paris on another peace mission, as his uncle wanted him to have an audience with Louis XIV to plead the case for peace talks to take place in Cologne. If this didn't work, Melchior was to go to London and appeal to Charles II.

Since Leibniz was already in Paris and had also worked for the elector of Mainz, he was enlisted to help Melchior. But when the day came to seek an audience with the king, Melchior alone was permitted to see Louis XIV and little came of the meeting.

At that time the French and English offensive in Holland had stalled. Leibniz and Melchior continued with their plan and sought to seize the opportunity to further peace efforts, by seeking an urgent consultation with the English court and presenting their proposal there. They set out in middle of winter for London, and arrived in Dover on January 21, 1673. It was almost exactly one year after Hooke had attacked Newton for his optics work.

In London, Leibniz's and Melchior's efforts to plead the case to the British king fell flat. And why wouldn't they? Charles II had agreed to join France in war and attack Holland. England and Holland had been at odds for years, and for his pains, Charles was rewarded with a yearly pension of £100,000 from Louis XIV.

However, this trip had a dual purpose for Leibniz. When in London, he also met with members of the Royal Society and made contact with some preeminent British scientists, particularly Robert Boyle, John Pell, and Robert Hooke, who discussed natural philosophy, mathematics, and chemistry with him. In this sense, London held as much excitement for Leibniz as Paris had. One figure Leibniz did not have the opportunity to meet, however, was Newton, who was in Cambridge at the time. Leibniz certainly would have been aware of him—as one of the brilliant young mathematicians who, like Leibniz, had just been elected to the Royal Society.

Leibniz had been aware of the Royal Society for a few years. In 1670, he had written a paper on the collision of bodies, called "A New Physical Hypothesis," in response to essays by Christopher Wren in England and Christian Huygens in France. The first part was on "concrete" motion, and the second on "abstract" motion. He dedicated the former to the Royal Society in London and the latter to the Académie des Sciences in Paris.

Academic societies were nothing new. Leibniz belonged to more than one when he was in college—but those were more informal than what emerged in Paris and London in the seventeenth century. The Académie des Sciences, for instance, was granted a royal charter and a room in the royal library at the Versailles Palace in 1666, and the moment when the charter was signed was regarded as such an important event that it was the subject of a painting by the artist Henri Testelin. In the painting, Louis XIV is depicted presenting the charter to a group of Académie founders.

In England, a group of churchmen, mathematicians, natural philosophers, and other scholars founded what would eventually become the Royal Society when they began meeting once a week in 1645 to "discourse upon such subjects" as natural and experimental philosophy. A number of individuals, including the mathematician

John Wallis, the astronomer Seth Ward, the chemist Robert Boyle, the statistical theorist William Petty, and the architect Christopher Wren attended these meetings, which were sometimes at a Dr. Jonathan Goddard's home, and sometimes at the lodgings of John Wilkins. When Wallis moved to Oxford as a professor a few years later, the group continued to meet in London and also began meeting in Oxford. The "Invisible College," as Boyle called it, was the home to lively discussions in math, physics, astronomy, architecture, magnetism, navigation, chemistry, and medicine—all the important subjects of the day.

The meetings continued on and off through the years when Newton and Leibniz were in school. When Oliver Cromwell died in 1658, the Invisible College stopped meeting due to the turmoil, but after the monarchy was restored and King Charles II came to the throne, the Invisible College was resurrected and reborn on July 15, 1662, as the Royal Society for London for Improving Natural Knowledge, with ninety-eight charter members. In the next twenty-five years, about three hundred new members were added, including Leibniz and Newton.

Part of the reason for the success of these societies was that science was becoming fashionable. There was great patronage of scientists among the wealthy and noble of Europe. Members of the Académie des Sciences received salaries from the government and funds for their experiments. High-society types attended chemistry lectures in Paris and London, and joined the Académie des Sciences and the Royal Society. King Charles II had his own chemical laboratory built, and aristocrats read scientific publications.

And what a sweet time of discovery the seventeenth century was. The diameter of the Earth was estimated to within a few yards, and a sophisticated modern view of the solar system evolved, with the orbits of heavenly bodies accurately tracked by telescopes and faithfully described by mathematics. The circulation of blood through the body was carefully charted, and microscopes led to the discovery of cells and a world of tiny organisms too small to be seen with the naked eye.

In 1673, when Leibniz was visiting the Royal Society, he was thinking of presenting an invention he had been working on for a while in Paris—a mechanical calculating machine, which Huygens called "a promising project" in a letter to Henry Oldenburg, the secretary of the still new Royal Society. As a friend of Boineburg's and fellow countryman, Oldenburg not only knew who Leibniz was, but had been in correspondence with him for a several years. Oldenburg was committed to helping Leibniz, who expected to make a splash in London with his calculating machine.

The Royal Society extended an invitation to Leibniz to demonstrate the machine. This wooden and metal device used a mechanical wheel to manipulate numbers. The famous French mathematician Blaise Pascal had invented a similar machine that could add and subtract, but Leibniz's could add, subtract, multiply, and divide. Or at least it was supposed to. In 1673, his calculating machine was an incomplete, nonfunctioning prototype when Leibniz hauled it across the English Channel. Leibniz's machine was something of a flop because he chose to demonstrate it even though it was not finished. He could explain it all very well, but his demonstration must have been something like the traveling vacuum cleaner salesman trying to sell his goods door-to-door during a blackout. The machine is great, and it would be very useful if only the darned thing worked.

Particularly unimpressed was Robert Hooke, who was spoiling for a fight. In addition to being one of the most prolific minds in the seventeenth century, Hooke was great with his hands and had made many scientific instruments; produced works of importance in astronomy, physics, biology; proposed a wave theory of light; discovered a new star in the constellation Orion; proposed the kinetic theory of gasses; and is still famous today for the discovery of the law governing masses on springs that bears his name.

Hooke was the toast of the Royal Society when Leibniz came to demonstrate his unfinished calculator in 1673, and, as Newton had already discovered, Hooke was infamous as for engaging in brutal disputes—not always within the boundaries of fair and open scientific debate—with his rivals. One example is Hooke's reaction to the spring balance, which Huygens had discovered as a by-product of his work in the 1650s, while trying to make a pendulum clock. Hooke not only disputed Huygens's discovery, he claimed it as his own, constructing a pocket watch and presenting it to England's king in the summer of 1675. Hooke went so far as to accuse Oldenburg, the secretary of the Royal Society, of betraying his ideas to Huygens.

Hooke clashed equally fiercely with Leibniz over the calculating machine. After looking carefully at all sides of it, and examining it in detail on February 1, 1673, Hooke expressed a desire to take it apart completely to examine its insides. This is no surprise—the machine is a tempting object for the curious.

In Hanover, there is a replica of Leibniz's machine on display. It's a fascinating object. Eight dials across the top allow a user to dial in a number, and add or subtract the numbers, which would reset the dials. The machine would keep track of the accumulating sum or difference. A knob on the machine acts to multiply or divide. Crank the handle one way and it divides, turn it the other way and it multiplies. The machine has a row of pentagons to address the problem of incorrectly adding columns of numbers with a different number of significant digits. The curator who put the machine on display had the foresight to place it on a mirror so that, by peering over it and peeking around it, one can examine nearly every inch of it. It's a fascinating machine and I can easily imagine how much Hooke wanted to take it apart.

A few days after Leibniz's presentation, Hooke attacked him in public, making derogatory comments about the machine and promising to construct his own superior and working calculating machine, which he would present to the society. At the same meeting, Hooke attacked Newton, lambasting him in a letter he read in front of the entire assembled Royal Society. Neither Newton nor Leibniz were not there to defend themselves, and Leibniz had to hear about the attack from Oldenburg, who assured him that Hooke was quarrelsome and cantankerous, and urged him that his best course of action would be to finish his machine as quickly as possible.

Hooke finished his machine, based upon the designs of his countryman Samuel Morland, and presented it on March 5, 1673, as promised. This must have made Leibniz's machine seem that much more unsatisfactory. Hooke made his in a matter of a few days, after all, and it worked as he said it would. Leibniz had been working on his machine for untold months, and so far it couldn't be shown to do anything.

Hooke's attack notwithstanding, the Royal Society later elected Leibniz a fellow on April 19, 1673, with the backing of Oldenburg. Leibniz committed a social faux pas by not immediately sending the Royal Society a formal letter of acceptance, as was the fashion of the times. Instead he sent a short note of thanks a few weeks later, which caused some grumblings among the fellows at the Royal Society. Oldenburg had to inform Leibniz that he was expected to write the formal letter, which he finally did several weeks later.

But worse embarrassment was yet to come for Leibniz after he visited Robert Boyle on February 12—the occasion of an event I like to call "The affair of the eyebrow."

Leibniz had been happy to meet Boyle, the long, gaunt older scientist, because he was interested in his experiments—and for good reason. Boyle, one of the founders of the Royal Society, was a brilliant experimentalist and was given to stunning audiences with his scientific demonstrations—such as when he proved that sound was carried by air by enclosing a bell within a glass jar from which he could remove all the air with a vacuum pump. When he rang the bell with the air removed from the jar, it made no noise whatsoever. Boyle also did carefully controlled experiments designed to demonstrate relationships between factors such as the pressure and volume of gasses or reactions between two compounds. He discovered the fact that certain vegetable abstracts change color when subjected to acids or bases—the technology behind litmus tests. And finally, he published his book, *The Skeptical Chymist*, in 1661. In this he abandons air, fire, water, and air as the elements and argues that the real elements are more primitive and simple.

The affair of the eyebrow began at Robert Boyle's house, when Leibniz met John Pell. Pell is a somewhat obscure figure today, but at the time was considered one of the top two or three mathematicians in Britain, an accomplishment made all the more remarkable by the fact that his reputation seems to have surpassed his actual work. But then Pell survived on his reputation alone. He had been a diplomat under Cromwell, stationed in Switzerland, so it is no wonder that when Charles II returned and Cromwell's head was placed on a spit above the streets of London, Pell's political career ended. Leibniz met him after these events.

Still, Pell was an expert on the sort of mathematics that Leibniz had been working on in Paris, in fact the type that Leibniz was presenting that night. Leibniz had arranged to have some of the work he had done in Paris written out, and he brought it with him to London so that he could present it anyone who was interested. At Boyle's house, Leibniz tried to impress the company by telling them that he had an original mathematical method for performing a difficult algebraic trick—employing the subtractions of square roots.

After looking at some of these "original" discoveries, Pell informed Leibniz that, a few years earlier, another mathematician, Gabriel Mouton, had published the same results in a book about the diameter of the sun and moon: *Observationes diametrorum solis et lunæ apparentium*. Mouton had reported in his book the results of a French mathematician, François Regnauld, in which Leibniz's supposed original discoveries had already been described. The same night Pell told Leibniz about the book, Leibniz grabbed a copy of it from Oldenburg, who lived close by. He opened it up and discovered that Pell was absolutely right. What an embarrassment. The book was available in France, and even though Leibniz had never heard of the book, there was the possibility that he might have read it.

This caused some, no doubt, to raise an eyebrow. Had Leibniz borrowed his ideas? Was he a plagiarist? Oldenburg asked him to write an explanation and deposit it in the papers of the Royal Society, which he did in haste. The letter that he wrote explaining the whole event was to become one of the key documents in the calculus wars. Even though it seemed like a simple misunderstanding, the letter proved that there had been a controversy—the possibility that Leibniz had plagiarized before. And for this reason it was an important document. Newton had a copy of it, apparently, in his possession when he died.

The affair of the eyebrow was a painful episode, but it revealed to Leibniz exactly how much mathematics—or rather how little—he understood, and he was left somewhat shaken at this humbling realization. At the very end of his life, Leibniz reflected on his lack of knowledge when he was visiting London. "Mathematics were studied by me only incidentally," he admitted. "I had not the least knowledge of the infinite series of Mercator; and as little of the advancement then made in the science of geometry, by the adoption of the new methods of investigation," he wrote. "I was not even thoroughly versed in the analysis of Descartes."

He would soon have the opportunity to know the works of Descartes and many others quite well. Though the affair of the eyebrow gave him a certain amount of grief, that grief gave him resolve to redouble his efforts to learn mathematics, and he would soon have ample opportunity to do so. The same night he visited Boyle, February 12, 1673, Johann Philipp von Schönborn, the elector of Mainz, died. Shortly thereafter, Leibniz and Melchior received the news and rushed back to Paris. Melchior went on to Germany to be close to the new prince, to whom he was related and who appointed him to the new court.

Leibniz left a letter for Oldenburg before he departed, requesting membership in the Royal Society, and he sent Oldenburg several letters from Paris in 1673—one in March, another in April, another in May, again in June and July, and another in October. Then he stopped writing for a time. Back in Paris, he redoubled his efforts to learn mathematics. The affair of the eyebrow showed him how much work he had yet to do. In this sense, Leibniz was not led to calculus so much as he was driven from a mixture of ambition and embarrassment.

For Newton, already in possession of publishable material on calculus, there was no running back to Paris or ignoring correspondence. He continued to correspond with Hooke and others on his theory of colors, and the effect this had was to make him crawl further and further away from any possibility that he would publish his mathematical work.

Oldenburg and Collins had given Leibniz a letter to deliver to Huygens upon his return to Paris, and when he delivered it, Huygens gave him many suggestions on what he could read. Leibniz was reading Barrow's book, which he had purchased in London, at the time and followed the meeting by seeking out the works of all the important mathematicians of his day—buying copies where he could, borrowing others, and transcribing information by hand. He read, absorbed, and sought the common threads in everything, and he made tremendous strides in the coming months.

Leibniz read the works of René Descartes, who had been such a profoundly important mathematician a generation before, and was even privy to some of Descartes' unpublished writings. Leibniz read Bonaventura Cavalieri's 1635 book, *Geometria*, in which he had developed new ways of analyzing geometrical shapes—a method of finding areas and volumes of geometrical shapes, which could be considered precursor work to calculus. Leibniz read Evangelista Torricelli, who developed methods for finding areas under parabolic curves and rendered a clear explanation of them. He read Gilles Personne de Roberval and Blaise Pascal, whose work on indivisibles and infinitesimals anticipated integral calculus. Leibniz knew of Johann Hudde, who in 1659 had given his own rule for constructing tangents and for geometrically finding the maxima and minima of algebraic equations. And he read René François de Sluse, who had made a rule for constructing tangents to a point on a curve.

He had an incredible propensity toward mathematics, and his lack of formal training in the subject probably helped him in the long run, contributing to the originality of his work (though it may have hurt him in the long run as well, since his lack of training also predisposed him to making errors). Mistakes aside, by the end of 1673, Leibniz had developed a way to use a series of rational numbers to find the solution to a problem that had vexed his contemporaries for a few years—the squaring of a circle, or a square equal to the area of a circle. Huygens described Leibniz's solution as being "very beautiful and very successful."

That was not all. Leibniz realized that Pascal's work could be combined with Sluse's tangent rule and applied to any geometrical curve, not just a circle. That is what led him to calculus.

5

Farewell and Think Kindly of Me

■ **1 6 7 3 – 1 6 7 7** ■

*It is an extremely useful thing to have knowledge of the true origins
of memorable discoveries, especially those that have been found not be
accident but by dint of meditation . . . the art of making discoveries
should be extended by considering noteworthy examples of it.*

—Leibniz, *History and Origin of the Differential Calculus,* 1714

In Paris, Leibniz still had to worry about his career, which was
suddenly uncertain, and he began making inquiries about
other jobs. The death of the elector presented problems for Leibniz,
in that he was owed two years of back pay from the old elector. He
enlisted young Schönborn to ask the new elector's permission to
remain in Paris, become a political emissary, and report on the political, scientific, and cultural events that were taking place. The
response, which eventually came, was that he could stay "for a while"
and keep his position as counselor, but he would receive no salary
and would not be promoted to emissary.

Things were far from desperate, however, because before he died,
Boineburg had made arrangements to send his son, who was a few

years younger than Leibniz, to Paris so that he could study under Leibniz. Thus, Leibniz continued to be employed by the estate for over a year, tutoring Boineburg's son, Philip William, who arrived in Paris on November 5, 1672. But Boineburg's son clashed with his tutor—playboy aristocrat versus the solitary genius. When young Philip William grew up, he would become a famous governor, be elevated to the noble rank of count, and became known as the "Great Boineburg." But in the 1670s, Philip William had no inclination toward serious study, especially not of the sort that Leibniz envisioned—a program that was to last from 6:00 A.M. until 10:00 P.M. The seventeen-year-old, a noble from one of Europe's boon docks, was in the prime of his youth and set loose on the decadent courts of Paris. He preferred to spend his time with his friends, and this caused friction between Leibniz and Philip William. As one nineteenth-century account put it, the young baron was smart and talented, but he was of an age in which he "manifested at that time a greater fondness for the sports which invigorated the body, than for the severe studies designed to develop the mind."

Leibniz wrote a letter to the Boineburg family complaining about his charge and asking for money to cover his expenses for tutoring and for his previous work on behalf of the boy's now-deceased father. In response, early in 1673, the boy's mother ended the tutoring and reduced Leibniz's pay. Leibniz was dismissed coldly from the employ of the Boineburg estate on September 13, 1674.

Leibniz now sought other employment. Through his friend Christian Habbeus von Lichtenstern, he was offered a position as secretary to the chief minister of the king of Denmark. This, he politely refused. Leibniz desperately wanted to stay in Paris and, from 1673 to 1676, sought continuously to secure a diplomatic or academic position that would allow him to stay there. Unfortunately, the fact that he was not of noble birth was a deal killer for his lofty diplomatic ambitions. Despite his brilliance, despite his charm, and despite his command of seventeenth-century law, he was of little use in diplomacy.

He also tried to obtain a salaried position at the Paris Académie des Sciences—similar to the position his mentor Huygens enjoyed. As a foreigner, such a position was not easily obtained. The fact that the Académie des Sciences paid salaries to its members meant that there was additional scrutiny over who should and who should not be a member. And, like almost everything else in seventeenth-century France, this question was clouded by nationalistic pride. French members of the academy, apparently, felt that there were already enough foreigners in the organization, and that the position and money should properly go to another Frenchman.

Huygens, the most prominent foreigner in the Académie des Sciences, could have helped Leibniz, but he was too busy and distracted at the time. Leibniz tried unsuccessfully to get an audience with the French minister Colbert for help, but failed.

So he tried other ways to gain entry. In typical fashion of the seventeenth-century French society, securing one of these coveted positions required currying favor with important individuals, and this meant having to make bribes. Leibniz was willing to try anything, and he befriended the Abbé Gallois, a man who made up for what he lacked in intelligence with his ability to climb the social ladder. Gallois could have helped arrange for a position for him but, unfortunately, these designs went bust after Leibniz snickered during a presentation Gallois made regarding the war in Holland. The Frenchman was greatly offended and immediately dropped support for Leibniz's cause.

In the end, Leibniz was forced to accept what perhaps was not his first choice of occupations: working for Duke Johann Friedrich of Hanover, a position he was offered on April 25, 1673. Leibniz had come to the duke's attention a few years earlier, and Johann Friedrich had invited him to Hanover then, but Leibniz had declined at the time since things were going so well in Mainz. He did, however, continue to correspond with the duke for the next few years. In 1671, for instance, he sent him two original papers, "On the Utility and Necessity of Demonstrating the Immortality of the Soul" and "On the Resurrection of Bodies." Leibniz also sent him a letter with an account of his research in multiple fields, including his idea for making the alphabet of human thought—giving himself an intellectual CV of sorts.

After the affair of the eyebrow, and the deaths of Boineburg and the elector of Mainz, Leibniz—now on the job market—wrote to Johann Friedrich almost as soon as he arrived back in Paris. To the duke, it was not a subtle hint, and Johann Friedrich jumped at the opportunity of bringing him to his court. He wrote back offering a position with a salary, and, to sweeten the deal, did not demand that Leibniz return from Paris immediately.

For Leibniz, this was a sweet deal because he had no desire to leave Paris. In fact, even after he accepted the duke's offer, he strung Johann Friedrich along for years, establishing the terms of the office, asking for more time to finish his calculating machine, asking to finish his mathematical research, and negotiating other matters with him. To the duke, Leibniz was boastful to a fault about his calculating machine, saying that it was regarded in both Paris and London as one of the great inventions of the time. He wrote to the Johann Friedrich on January 21, 1675, asking if he wanted one of the calculating machines constructed for him.

Leibniz had set about supervising the completion of his calculating machine as soon as he returned from London. Always the optimist, Leibniz told Henry Oldenburg that he expected to be done very soon. But he was ultimately not satisfied with the design and decided to make radical revisions. Then, when the design was done, and the machine all but built, the craftsmen working on the project for Leibniz lost interest. Leibniz delayed writing to Oldenburg for months and months, Back in England, Oldenburg was probably wondering what had happened to him. It had been over a year since their last correspondence. Finally, in the fall of 1674, Leibniz had a Danish nobleman, Christian Walter, who was going to England, hand-deliver a letter to Oldenburg. In it, he said that his calculating machine was finally finished, and claimed that it could multiply a ten-figure number by a four-figure number, with but four turns of the crank to get the answer.

Once the machine was finished, Leibniz invited scientists to his rooms in Paris and demonstrated it—to the apparent wonderment of those who witnessed it. One of the people who came was Étienne Périer, who was the nephew of Blaise Pascal, who had invented the precursor machine in Paris some twenty years before. Leibniz's machine was a vast improvement of Pascal's machine, which was only able to add and subtract, as it added the two other fundamental algebraic operations, multiplication and division.

Leibniz was a larger-than-life figure, gangly, with long fingers and limbs, and a huge wig and courtly clothes. It's easy to imagine him, with sweeping gestures, describing the uses of the machine: a marvelous speaker, now he's talking about how addition and subtraction only required a few turns of the wheel. Whole pages of numbers can be added and subtracted faster than it would take to even write them down. Now he's on to multiplication and division. The French finance minister, Colbert, wanted three—one for the king, one for the Royal Observatory, and one for his own financial offices.

Leibniz's calculating machine was only a small part of what its inventor was doing during this time. He also threw himself into mathematical studies, teaching himself much of seventeenth-century mathematics within a few years. In fact, when Leibniz wrote to Oldenburg in the fall of 1674, after more than a year of silence, it was not about his model of the calculating machine but rather about some mathematical work he had been doing. By 1674, after more than a year of exhaustive work, Leibniz had arrived at the same place Newton had independently reached just a few years before. Leibniz still knew very little of the work of Newton, but that was about to change, thanks to Oldenburg.

Oldenburg is practically the inventor of modern scientific discourse—not because he developed any fundamental technology or pilloried the scientific journals with his papers, but because he was behind the success of what was really the first successful scientific journal—the *Philosophical Transactions of the Royal Society*. Oldenburg was the founding editor of the *Philosophical Transactions*, which he launched on July 3, 1665, and supervised until issue number 136 in June 1677.

The story of how Oldenburg came to play such an important role in the Royal Society is an interesting one. He was born in Bremen and came to England in 1653 as Bremin's London consul during the reign of Cromwell. He lost his job a few years later and became a private tutor for a British nobleman's children in London; when they moved to Oxford in 1656, their tutor moved with them. This was fortuitous for Oldenburg because in Oxford he made the acquaintance of those philosophers who would come together and form the Royal Society.

He was one of the first members of the Royal Society, and he was the secretary of the Royal Society from 1663 until his death. During the nearly fifteen years that he held that position, he was one of the most important members of the society. A prolific letter writer, he kept a correspondence with more than seventy philosophers and mathematicians. Many of these letters were written to communicate discoveries between various philosophers, mathematicians, and scientists throughout Europe. In addition to serving as secretary and furthering the science of British mathematicians through the publication of the *Philosophical Transactions*, he welcomed the cream of contemporary continental scientists into the society—men like the French astronomer Giovanni Cassini, the Dutch physicist and mathematician Christian Huygens, the Italian doctor and anatomist Marcello Malpighi, the early microbiologist Antoni van Leeuwenhoek, and of course Leibniz.

He kept so much correspondence, in fact, that he drew suspicion of certain officials and was arrested "for dangerous designs and practices"; he was locked in the Tower of London in the summer of 1667 but released after two months. Oldenburg actually deserves a great deal more credit than such biased suspicion afforded him. For the last few years of his life, if there was a discovery being made in England or the continent, he was probably in the middle of communicating it.

He was also involved in disputes between various people, such as when Huygens became embroiled in his fight with Hooke after he invented his balance spring, a device that uses oscillations to regulate the movement of a clock. It was a significant technological improvement at the time, and Huygens sought and was granted a patent for it by Colbert, the French minister of finance. Huygens also registered his invention of the balance spring with the British, in a manner of speaking, by sending to the Royal Society a letter containing a coded anagram description of it. Later, he sent a full description, and when this description was read at a meeting of the Royal Society on February 18, 1675, Hooke lashed out at Oldenburg, claiming that he had inverted the balance spring himself, accusing Oldenburg of spilling the beans to Huygens, and suggesting the venerable secretary was a French spy. The Royal Society backed Oldenburg against Hooke's claims, but these charges would unfortunately linger over his head long after he died—complicated no doubt by the central role that he played in the calculus wars by fostering communication between Newton and Leibniz.

Twenty years after he first came to England, Oldenburg was perhaps the only person alive who was in continuous contact with Newton and Leibniz for the entire time the latter was in Paris. He enabled their first correspondence—two letters each, which were written by one, passed to Oldenburg, and then forwarded to the other.

What led to this exchange of letters was the correspondence Oldenburg himself carried on with Leibniz after Leibniz returned to Paris from London. They had been occasional correspondents for a few years before the two finally met in 1673, while Leibniz was visiting London, and afterward the two were in close contact. Oldenburg had taken an interest in Leibniz as his fellow countryman and a brilliant thinker. Leibniz had mutual admiration for the older German, since he was a friend of Boineburg's, and believed Oldenburg would be a good source of information for him on the state of mathematical discoveries in Britain. And Leibniz was right. Oldenburg did as much as he could to share with him information about the state of British mathematics.

This mutual exchange would lead some to believe the sort of accusation that Hooke made—that, in fact, Oldenburg was some sort of spy. In fact, one nineteenth-century account of the calculus wars makes quite a lot of the fact that Oldenburg and Leibniz were both from northern Germany. "The Royal Society in London had committed the oversight of employing as their secretary, not an Englishman, but a German, Heinrich Oldenburg," the writer, a Dr. H. Sloman, said. "This imprudence could not but soon have its consequence, and this consequence in particular, that when once the right man came, the interest of England was more or less sacrificed to a German friendship."

Sloman's book claimed Oldenburg promoted a young and overly ambitious Leibniz, who took advantage of Oldenburg's natural bias toward his fellow countryman and made Oldenburg his "agent." Sloman had Oldenburg conspiring with Leibniz and shuffling the younger man through a side door of the Royal Society into a prestigious membership on nothing more than the older German's assurance of Leibniz's genius and not on the merits of his nominee's work.

"Oldenburg here again contrives his defense," Sloman wrote of the secretary's reaction to the affair of the eyebrow. "And as Leibniz had now become his pet and favorite, he exerted himself for his fame more than for his own . . . and so we see with astonishment the endeavors of the two friends quickly crowned in the access of the young man in the honor of becoming a member of the Royal Society."

This is ludicrous for a few reasons, not the least of which is the fact that there were contemporary members of the Royal Society who were far less accomplished than Leibniz—even at that early age. Nevertheless, there is no question that Oldenburg's communications with Leibniz did more to fan the flames of the calculus wars years later, when it exploded after the turn of the eighteenth century, than nearly anything else that happened in the 1670s.

A critical exchange took place in April 1673, when Leibniz received a long letter from Oldenburg. In the early 1670s, Oldenburg was compiling a roundup of all the great accomplishments of British mathematicians based on information he was gathering from others, particularly John Collins, who has been described as a pygmy standing between two giants. He was a minor government servant, an accountant—a mathematical hobbyist really—who, by luck, chanced to be central to one of the only exchanges of letters between the two greatest mathematical geniuses alive in his day.

The son of a poor preacher outside Oxford, Collins was a bookseller's apprentice who later spent seven years as a seaman in service against the Ottoman Empire. Later, he became a mathematics teacher, an accountant, and finally (owing to the fact that he was a likable chap) a well-connected mathematician. Although he never contributed great mathematical discoveries like Newton and Leibniz, nor was he a consummate enabler of correspondence like Oldenburg, he nevertheless knew enough to comment on the work of others, and he could recognize great work when he saw it. Because he understood algebra, Collins was involved in Oldenburg's communications with Newton and Leibniz. Oldenburg was not himself a mathematician, and could do little with obscure mathematical discoveries entrusted to him without help.

Collins was in a perfect position to be that help. He was one of the few who were privy to Newton's early results as a mathematician. Newton had written letters to Collins in the early 1670s describing a number of his results, and they had carried on a lively correspondence for several years. Collins was happy to communicate these results to Oldenburg because he was what one might call a mathematical anglophile—one who wasted no opportunity to assert British superiority in math or science.

In his position as mathematical intermediary, Collins helped Oldenburg to draft Leibniz a letter detailing the status of mathematics in Britain—including the work of Newton. For Leibniz, the most valuable part of the letter was probably references to the contemporary British publications that Collins had meticulously compiled. This report contained references to books and papers that revealed to Leibniz the existence of a whole literature of mathematics that he scarcely knew existed.

The mathematical details sent to Leibniz were purposely vague, though, because Collins was cautious about revealing too much about his countrymen's proprietary discoveries. He regarded the French with particular suspicion, and though Leibniz was not French, he carried the stain of living in Paris. Plus, the young German was a protégé of Huygens, who was then seen as one of the main competitors of English mathematicians.

So as much as Collins revealed, he withheld. To Leibniz, he described results of work by Newton and the Scottish mathematician James Gregory on infinitesimals, for instance, listing problems that Newton and Gregory could solve—but not their methods. This vagueness was unfortunate because it later led Leibniz to believe that his growth in mathematical discovery was completely fresh. There were plenty of other mathematicians who had solved the sort of problems calculus could solve, by using methods other than calculus. Leibniz would think that he was making completely original strides while, in fact, much of what he was discovering had already largely been worked out by Newton; it just hadn't been published— partly because of the Great Fire of London and partly because of Newton's trouble with Hooke.

An example of the level of detail, or rather the lack thereof, can be appreciated in the following passage:

As to solid or curvilinear geometry, Mr. Newton hath invented (before Mercator publish't his Logarithmotechnia) a general method of the same kind for the quadrature of all curvilinear figures, the straightening of curves, the finding of the centers of gravity and solidity of all round solids and of their second segments ... which doctrine, I hope, Mr. Newton is a publishing. ...

After receiving this letter, Leibniz went more than a year without writing anything to Oldenburg at the Royal Society. Following up on the references Collins provided, Leibniz was astounded to find out that, besides the material he presented to Pell that caused the affair of the eyebrow, much more of the mathematical work he was doing had already been done by others. Astounded and excited at the same time, he now knew what he didn't know. Leibniz withdrew into the cell of his mind and began to work and rework the mathematics that he had to understand.

When Leibniz wrote to Oldenburg in the summer of 1674, after his many months of silence, Oldenburg had no way of knowing what an expert mathematician Leibniz had become, but that is how Leibniz presented himself in his letter, "In geometry, I have made some discoveries by rare luck ... theorems of greater importance [including] certain analytical methods, completely general and widely extended, which I value more highly than particular theorems however excellent." And as if too excited to wait for a reply, Leibniz wrote another letter a few weeks later, reiterating that he had made "a notable discovery" in the branch of geometry involving the analysis of curves.

Oldenburg replied on December 8, 1674, that Newton and Gregory both had general methods for all geometrical curves by which they could determine surface areas and volumes and other functions related to curves, such as tangents. Leibniz wrote Oldenburg yet again on March 30, 1675, excited about Newton and his work. "You write that your distinguished Newton has a method of expressing all squarings, and the measures of all curves, surfaces and solids generated by revolution, as well as the finding of centres of gravity, by a method of approximations of course, for this is what I infer it to be. Such a method, if it is universal and convenient, deserves to be appraised, and I have no doubt that it will prove worthy of its most brilliant discoverer."

Thus began the exchange of letters involving Leibniz, Oldenburg, Collins, and eventually Newton in the last two years Leibniz was in Paris. They corresponded more or less continuously, playing a sort of cat-and-mouse game, with Leibniz sharing information, holding some back, and Collins doing the same thing. Leibniz began asking a number of questions about a specific type of geometrical problem called a quadrature. Quadratures were one of the hot topics in the 1670s, and many different mathematicians were working on difficult solutions to them. Calculus makes solving quadrature problems trivial. He also began to boast about his own methods—if in the most vague possible terms, taking his cue from the previous exchange with Collins.

He began asking specific questions about Newton and Gregory's results—did they have methods for rectifying the hyperbola and the ellipse? He offered to trade his own "far reaching" methods for some of Newton and Gregory's methods that he knew Collins possessed. Leibniz was now very interested in what Newton had to offer, for it seemed to him that Newton had already made a lot of progress in this area.

Meanwhile, Leibniz made superb progress on his own. He had gotten a good start in mathematics thanks to his study of the work that Pascal and others had already done, and soon he began to make important discoveries of his own. One was a technique he called the transmutation rule, which was a way of figuring the quadrature of a curve, an important step along his way to inventing calculus.

By October 1675, having absorbed everything he could from his contemporaries, pulling together their work in his self-imposed retreat, he came out of his intellectual gestation and forged ahead. In 1675, Leibniz moved beyond the body of available knowledge and into the uncharted territory of differential calculus. In October and November of that year, he was able to bring these ideas together in a number of notes and papers he wrote containing the essence of calculus.

Moreover, Leibniz invented the symbols of differential and integral calculus, as we know them today. On October 29, for instance, he came up with the integral sign. Leibniz saw integration as summation. In fact, that's why he gave it his symbol, "$\int$," which is a fancy S that he invented. The new symbolism provided a general way to treat infinitesimal problems of calculus and would prove most useful for its spread.

This was a notion that appealed completely to Leibniz, who always favored utilitarian ends. Even in his younger days, when he was a mathematical novice, he was keen on communications being easily understandable. For instance, he praised the work of a philosopher, Nizolius, not for his philosophy itself, which contained many errors, in his opinion, but for its clear literary style. Nizolius, in fact, had suggested that anything that could not be described using simple terms expressed in everyday language was useless. In response to Nizolius, Leibniz recommended that jargon be avoided. In fact, one of his first introductions to mathematics while he was still in college was by a Professor Erhard Weigel, who had a reputation for taking apart other academics by asking them to repeat their Latin arguments in plain German. Weigel instilled in Leibniz a love for simplicity in discourse.

It's no surprise, then, that following his discoveries in calculus, Leibniz saw the need for a clear way of describing them. In creating a clear and compact language for his work, he became a master mathematician. Leibniz proved this soon after, when a French mathematician, Claude Milliet Deschales, asked him to determine what the part of a circular cone would be if you cut off the tip with a plane parallel to the base, and Leibniz was able to work this out in a single evening.

In the next couple of years, Leibniz developed his methods of calculus, but he wouldn't publish his work for another decade, which is something that is worthy of a comment.

Of all the nuanced differences between the work of a seventeenth-century scientist and a more modern one, none seems more pronounced than publishing. Today, publishing plays a central role in science and is integral to the advancement of nearly every scientist's career. In fact, research findings are not finished in a sense until they are published in a peer-reviewed journal, and scientists make their reputations based on the number and quality of such publications. Competition among scientists is fierce, and there is often a rush to publish discoveries almost as soon as they can be written and reviewed. In recent years, scientific journals have even taken to publishing papers online as soon as they are ready—and in some cases even before they are edited. Today the idea of not publishing and jealously guarding a work as profoundly important as calculus is foreign. Today's successful scientist making an original discovery that is worthy of publication will likely move with great speed to publish the results.

Leibniz might have published his calculus work earlier than he did, but the problem was that he needed to deal with much more pressing matters. As towering an achievement in the history of mathematics as inventing calculus as a relative novice may have been, it did little at the time to advance his career. His formal appointment to the Court of Hanover took place at the beginning of 1676, and, from that moment, the clock was ticking—the forces pulling him away from Paris were growing. At the end of February of that year, he was told that his patron duke wanted him to come to Hanover, and he would soon have to do so.

His uncertain future aside, Leibniz continued to work, correspond, and study. He wrote to his acquaintances on subjects that included law, gravity, and the logical underpinnings of experimental physics, and he had the desire to correspond on mathematics. He wrote to Oldenburg at the end of 1675, promising to show the solution to an unsolved problem in geometry that he had solved using new methods he had invented—an allusion to calculus.

Now the stage was set. These discoveries that Leibniz made in the waning months of 1675 would bring him, within a year, in contact with Newton. Just before he left Paris, Leibniz and Newton exchanged a few letters in which they danced around the subject of calculus. Their exchange had all the outward dull politeness of academic courtesy, and there is little there that anticipates the polemics that they wrote about each other decades later, when the calculus wars were at their climax.

Newton knew vaguely of Leibniz before their exchange, since he was familiar with one of Leibniz's fellow Germans, Ehrenfried Walther von Tschirnhaus, who arrived in Paris from Saxony in August 1675. Tschirnhaus soon became friends with Leibniz, and the two made a few joint studies in addition to having many mathematical discussions (in which Leibniz was clearly the master). But Newton was not impressed with Tschirnhaus, and by extension probably wasn't impressed with his countryman Leibniz.

At the time Leibniz was inventing calculus, Newton was still dealing with the fallout of publishing his theory of colors and was *still* having problems with his optical theories. He had been defending himself against Hooke and Huygens for more than three years, and the affront would not go away any time soon. Newton sent Oldenburg a long letter enclosed with a document, "An Hypothesis explaining the Properties of Light Discoursed of in my Several Papers," on December 7, 1675. The "Hypothesis" was an extensive defense of his optical theories.

Also in 1675, Newton made a trip to the Royal Society to attend a meeting—his first, even though he had been a member for three years. But far from striding triumphant into the hallowed halls, he was ready to cut himself off from communication just about the time when he would engage in the most important correspondence of the calculus wars. In fact, in five months' time, Hooke renewed his attack on Newton in early May 1676, by standing up and declaring at a meeting of the society that Newton's work on light was lifted from his own work, *Micrographia*. On May 25, a battered, agitated, and distracted Newton was approached by Collins and Oldenburg, and asked to write a letter to Leibniz. Newton was so embroiled with his battles over his optical work that he had little taste for opening himself up to a potential attack by revealing his work to a rival mathematician.

Still, Collins cajoled him to write to Leibniz because he was afraid that Leibniz was catching up with Newton. Collins was right. Leibniz was fast becoming every bit as brilliant the mathematician that Newton had been for a decade. Collins wasn't in the greatest of positions to carry on a correspondence during these days. He was at the end of his life and not in the best of health. And, in 1676, he lost his job. Nevertheless, that May, Collins heard from Oldenburg that

Leibniz was interested in further communication, and he began putting together a large account of the discoveries of James Gregory, who had recently died. This fifty-page document was later called the *Historiola*, and was meant to be a summary of English achievements in mathematics over the previous several decades.

In France and elsewhere on the continent, Descartes was still revered for his mathematical work, and his supremacy in mathematics was often asserted. But Collins felt that the Brits had made significant progress beyond Descartes, and the "*Historiola*" was an attempt to document this. Collins wrote the "*Historiola*" as a means to inform rather than instruct. He was not so much interested in teaching the mathematics of the British mathematicians to Leibniz as in securing their rights as inventors, and so he simply expounded which mathematicians had solved which problems, without going into methods or proofs.

Oldenburg thought the paper was running too long at fifty pages, so he asked Collins to abridge it. He then translated the abridgment into Latin. This was unfortunate because, in transcribing this complicated document into Latin, certain errors were made.

In any case, Leibniz would soon be receiving his first, enticing letter from a paranoid, battered Newton in the summer of 1676. Newton finished his *epistola prior*, as he would later call this letter, on June 13, and he sent it on to Oldenburg, who received it on June 23 and read it at the Royal Society a few days later. Sensing that this letter was of some importance, Oldenburg took extra measures to ensure that it was preserved and that Leibniz would get his copy. He had the letter copied and sent it to Leibniz some six weeks later, along with extracts from the letters of Gregory.

Not trusting the regular post, Oldenburg gave the package to a man named Samuel König. König, a German mathematician, would be leaving London around the beginning of August and heading to Paris. The timing was perfect—always better to wait a few more days and have it hand-delivered, Oldenburg must have reasoned. Once he got to Paris, however, König couldn't find Leibniz, so he left the package at a local store, thinking the owner would soon see the German to conclude the relay. As it happened, the letter languished until Leibniz wandered by weeks later on August 24, 1676, and found it . . . what's this? A letter from England?!—a letter from Newton himself!

The first letter is eleven pages long and is a catalog of the Englishman's mathematical results, detailing several problems that Newton was able to solve with his methods. The centerpiece of this letter was Newton's binomial theorem, a highly original discovery whereby roots of an equation can be extracted and a calculation simplified. The letter hints at "certain further methods" that Newton did not then have the time to explain. There was nothing in the letter of the central problem—that calculus could be used to solve these same infinite series problems.

Newton was being cautious; he may have suspected Leibniz was playing a complicated ruse to get him to reveal his secrets—by pretending that he had secrets of his own. Thus there was nothing in the letter that was not already known to Leibniz in some form or another. Nothing. The only new item, in fact, was added by Oldenburg—another reminder to Leibniz that his promised calculating machine was long overdue. "I would really like you, a German and a member of the said society, to fulfill the promise you gave, and in that way relieve me as soon as possible of an anxiety on account of a fellow citizen which vexes me very much," Oldenburg wrote, concluding his cover to Newton's *epistola prior*. "Farewell again, and pardon this frankness of mine."

The fact that Newton did not send his methods would be an important point when the dispute raged decades later, because Leibniz would legitimately claim that he got nothing from the English and Newton as far as the methods of calculus were concerned. For all Leibniz knew, Newton had one method for solving a problem and he had another. In fact, Newton seemed to say the same thing himself in the opening to his *epistola prior* and was perfectly willing to acknowledge that Leibniz had something mathematically. "I have no doubt that he has discovered [speedy methods] . . . perhaps like our own if not even better," he wrote in the first letter.

Oldenburg warned Leibniz that, in getting the letter transcribed, mistakes may have crept into the version that he now held, but that they should not present a problem for its recipient. "Your shrewdness will correct any errors," the older German wrote in his cover letter.

Leibniz was blown away by the *epistola prior*. He immediately dashed off a reply for Oldenburg to give to Newton, commenting that the letter had "more numerous and more remarkable ideas about analysis then many thick volumes printed on these matters." He called Newton's series work to be worthy of the man who came up with the theory of colors and who invented the reflecting telescope.

In his response, Leibniz described his own mathematics and described an original discovery of his own, called his transmutation theorem, but withholding descriptions of his methods just as Newton had withheld his. He also included his arithmetical quadrature of the circle, as promised but, again, as was characteristic of this entire exchange, sent only the basic details, withholding the critical secrets that allowed him to solve it, feeling that since Newton gave only his results he needed to do so as well. On the other hand, he asked many questions, clearly intending to maintain the correspondence. Knowing that he would shortly be leaving for Germany, Leibniz wrote this reply after just three days, sending it on August 27, 1676. He ended the portion of the letter that was written for Oldenburg with a polite salutation: "Farewell and think kindly of one who is devoted to you."

Leibniz was so excited and rushed that his scrawled letter contained several mistakes and was written in a thick chicken scratch that was hard for Collins to copy over for Newton, and Collins amplified the sloppiness with his own mistakes in transcribing the letter. Significantly, the date on the cover of the letter was miscopied, so years later, when Newton was re-creating the chronology of that summer, he assumed that Leibniz received the letter shortly after he sent it in June. When Newton was poring back over this material, he incorrectly assumed that Leibniz had taken six weeks to reply— ample time to consider the material contained therein at great length. Years later, some of Newton's supporters would also seize upon the mistakes as proof that Leibniz did not know what he was doing as opposed to being symptomatic of Leibniz's haste—as was his excited tone.

When Newton received Leibniz's reply, many weeks had in fact passed, and because he assumed that Leibniz had taken this time to write his letter, Newton decided to do the same thing and took his own time to reply back—a tragedy, as it turns out, because after spending six weeks crafting his second letter to Leibniz, which he later called the *epistola posterior*, Newton sent it on November 3, 1676, but it was by then too late to send it to Leibniz in Paris. It did not reach Leibniz for nearly a year, because by the time Newton mailed it, he had already left Paris for the last time. When it finally did reach him, he was in Hanover.

While Newton was mulling over making his response, Leibniz, having delayed returning to Germany as long as he possibly could, could delay no longer. The duke, turning up the heat, wrote to him several times in the summer of 1676, again asking Leibniz to come to his new job as quickly as possible. Leibniz stalled for a few more months. He got another letter from Hanover in July, and the tone of this one was different. Its writer, a court official named Kahn, expressed genuine surprise that he had delayed so long, perhaps sensing that Leibniz was not going to come at all. But rather than admonishing him the letter sought to sweeten the deal, and Kahn offered that, in addition to his post as a counselor, Leibniz could also be in charge of Johann Friedrich's library.

Ah, books! The duke and his men knew exactly what they were doing by offering this to Leibniz. It was like offering an addict his favorite drug. In July, Leibniz was given his travel expenses from the Hanoverian ambassador in Paris and finally, on September 13, 1676, the duke put his foot down, writing that Leibniz could either come to Hanover or forget it.

Leibniz now had no choice. By the end of September, he had delayed leaving Paris for as long as he could. Within days, he was forced to leave Paris for good—riding out of town with the mail coach on October 4, 1676. He had come to the city a young man primarily interested in law and matters of state, knowing very little mathematics, and left four years later one of the top two or three mathematicians in Europe. (Today there is a street in Paris, the rue Leibniz, named in his honor.)

But he was still not yet on his way to Hanover. On his way there, Leibniz made a few stops. First, Calais, where the autumn storms blew against the docked boats for nearly a week until he was able to board one and set sail for England, where he arrived on October 18. He stayed in London for a little more than a week—several days that would shake his world forty-five years later and become the cornerstone of the claim that Leibniz had benefited from seeing Newton's early work.

In London, Leibniz met Oldenburg again and showed him, at long last, the calculating machine. This meeting was rather insignificant historically—the much more important meeting on this trip was when Leibniz finally met Collins. Collins was apparently much charmed by his young guest despite the fact that he spoke no German and only poor Latin, and Leibniz had only poor English. But Collins liked the young man and he allowed him to peruse his correspondences and papers and to have access to the books in his possession, including some unpublished works of Newton's. Collins was the Royal Society librarian in those days, and the society was still on recess for that week, so there was really no harm at all, he thought.

Leibniz looked at Newton's "*De Analysi*" and took notes from it. He also looked at the long "*Historiola*," which would become the subject of accusations against him decades later. Newton was convinced that Leibniz had the "*Historiola*" with him in Paris because there was a note on the cover asking him to return it when he was done. The note was, of course, referring to when he was done with the book while in London, where he spent a few hours over a few days looking at it. Newton assumed that Leibniz spent months studying it, as opposed to taking the quick look at it and making the few notes that he did.

Nevertheless, this seventeenth-century equivalent of a Post-it note became evidence for Newton and his supporters later that Leibniz had read the "*Historiola*" and other documents while in London. The "*Historiola*" detailed a lot of information about Gregory, Pell, and Newton, and, in particular, under the auspices of Collins, Leibniz saw a letter Newton had written that contained a detailed explanation of his rule for finding tangents—the slope of a curve at any given point—which would be something that Newton would claim Leibniz stole from him.

Collins tried to get Newton to publish his calculus, but Newton was too burned by the experience of publishing his theory of colors that he would not even consider it. "I could wish I could retract what has been done," he wrote to Collins on November 8, 1676, "but by that, I have learnt what's to my convenience, which is to let what I write lie by till I am out of the way."

Newton also reassured Collins that his methods were superior to Leibniz's. "As for the apprehension that Mr. Leibniz's method may be more general or more easy then mine, you will not find any such thing. . . . The advantage of the way I follow you may guess by the conclusions drawn from it which I have set down in my answer to Mr. Leibniz: though I have not said all there."

A few months after Leibniz left, Collins wrote to Newton about the German's visit, saying that they had discussed some things taken from letters written by Gregory. However, Collins did not mention that he had let Leibniz see Newton's papers—perhaps feeling guilty about showing him so much.

A few years later, Collins died without Newton ever realizing what he had shown Leibniz. Only decades later, long after Leibniz published his calculus papers, would Newton piece together what had transpired during that late autumn week in London, but even then imperfectly, of course, because he would draw far too much significance from the fact that Leibniz had read Collins's copy of "*De Analysi*." "*De Analysi*" was a crucial document because it and other pieces of evidence proved that Newton had invented calculus before

Leibniz. But they were not enough to prove that Leibniz had borrowed any ideas from Newton, so the evidence that Leibniz was in London and looked at these works was essential for establishing the possibility that the German had stolen his work from Newton.

In fact, Leibniz did take notes from *"De Analysi,"* but the notes themselves are not really on the formulation of calculus but on some of the other things that are contained in the book. Today, there is little argument over the fact that Newton and Leibniz did their work independently of one another, because the documentation exists in Leibniz's notes from October 1675—many months before he saw anything of Newton's.

But the conflict was still to come; Leibniz was on his way back to Germany where he would start a new life, and Newton, apparently, was losing his interest in mathematics, which he referred to as dry and barren. Instead, he was becoming interested in alchemy and other subjects.

Leibniz left London feeling good about having finished his obligation with Oldenburg and having opened up a new line of communication with Collins. He set out for Germany aboard the yacht of Prince Ruprecht von der Pfalz, whom he met in London. He sailed first to Rotterdam, writing a discourse on the subject of a universal language while he was waiting to set sail, and complaining in a letter to an acquaintance that he had nobody to talk to but sailors.

From there, he made his way to Amsterdam, where he met with a few notable people, including Johann Hudde, the mathematician who had independently discovered many of the precursor methods to calculus—such as finding tangents to curves and doing the quadrature of the hyperbole. Then he went on a short tour of the surrounding country, visiting Haarlem, Leiden, Delft, the Hague, and finally back to Amsterdam. He met Antoni van Leeuwenhoek, who was a fellow member of the Royal Society and is still famous today for his discovery of microorganisms. He had long conversations with Benedict Spinoza on philosophy and theology.

Finally, he left for Germany and arrived in Hanover at the very end of 1676. As Leibniz's time in Paris had come to an end, so too was the war that Leibniz had gone to Paris to prevent coming to an end. It would finally be over, with the Treaty of Nijmegen, in 1678. The treaty allowed Holland to remain intact, and as a concession to France, Louis XIV was allowed to keep the Lorraine. The preparations for this treaty took much time, and, even a year earlier, Leibniz had been busy writing documents supporting what would eventually be the peace conference when he got Newton's very old and well-traveled second letter with a note from Oldenburg in June 1677.

As noted, this letter would not reach him for nearly a year after Newton had dispatched it. Oldenburg wrote his cover letter to Leibniz on February 22, 1677, explaining that he "put off writing to you until now, because I did not want to endanger what I have at hand for transmission to you, including a letter from Newton as weighty in argument as it is copious in expression."

In the second letter, nineteen pages long, Newton was even more superlative with his praise: "Leibniz's method of obtaining convergent series is certainly extremely elegant and would sufficiently display the writer's genius even if he should write nothing else." Newton also now expressed an interest in seeing Leibniz's results. He wrote, "The letter of the most excellent Leibniz fully deserved of course that I should give it this more extended reply. And this time I wanted to write in greater detail because I did not believe that your more engaging pursuits should often be interrupted by me with this rather austere kind of writing."

If Newton's letter was warm on the surface, it was frozen in the middle. He was not particularly enthusiastic to carry on the correspondence. He gave a rich though veiled description of some of his most important mathematics, writing again about his series methods and on his discovery of the binomial theorem, and touching on his methods of fluxions (calculus) by showing three examples, tantalizing Leibniz by stating that he had arrived at "certain general theorems." Of course, he was not willing to part with anything of real substance, so he refrained from going into too much detail.

What detail he did divulge, he pined over. After he sent the letter to Oldenburg toward the end of 1676, Newton sent another letter just days afterward, asking him to make a few changes. "Two days since, I sent you an answer to M. Leibniz's excellent Letter. After it was gone, running my eyes over a transcript that I had made to be taken of it, I found some things which I could wish altered, & since I cannot now do it myself, I desire you would do it for me, before you send it away."

So careful was Newton that, when he did disclose an important statement of calculus, he did so in an unintelligible form. He sent it in the form of an anagram—a common device in those days for asserting priority while not revealing anything. "The foundation of these operations is evident enough," Newton wrote. "But because I cannot proceed with the explanation now, I have preferred to conceal it thus: 6accdoe13eff7i319n4o4qrr4s8t12ux. . . ."

These secrets were transposed encoded characters. Once it was transposed properly and translated into Latin (and then into English), the sentence read: "Given in an equation the fluents of any number of quantities, to find the fluxions and vice versa."

How hard would it have been for Leibniz to read these lines? Impossible. To give a flavor of the difficulty, imagine reading a single word thus coded, "coffeepots," and trying to decipher its meaning. A simple cipher would be to replace each letter in "coffeepots" with the proceeding letter of the alphabet; the word would become "dpggffqput"; then, transposing these letters randomly would give something like "fpgqpufdtg." The word "fpgqpufdtg" bears little resemblance to "coffeepots," and likewise the sentence that Newton wrote was unrecognizable.

Writing anagrams was not so unusual. Huygens wrote his own anagram one time to conceal his invention of the spring balance for his pocket watch. Likewise, Newton was using an anagram to evidence the fact that he was in possession of his method of fluxions yet clearly did not intend to share that method; he would have known that Leibniz had absolutely no way of decoding the anagram. Moreover, even if he had the key to decipher the code, Leibniz would not have been able to decode the anagrams because one of them wasn't even transcribed correctly in the copy that was sent to him.

Indecipherable bits of the letter aside, Leibniz was thrilled to receive it. He had been in the intellectual backwater of Hanover for several months and must have been going through withdrawal when he received the *epistola posterior*. He immediately responded to Newton and Oldenburg just days later, on June 11, 1677, in a letter full of praise and inquiry. He communicated the essence of his differential calculus, and he implored Newton for further correspondence. "I am enormously pleased that he has described by what path he happened on some of his really very elegant theorems," he wrote, and he wrote again a few months later, practically begging Newton to open the communication. Leibniz further asked Oldenburg to send him copies of the *Philosophical Transactions* and news of other discoveries in Britain.

Oldenburg replied to Leibniz on August 9, 1677, telling him that Newton was preoccupied and thus he shouldn't expect a reply right away. Newton never did reply. Overtired from the dispute over his theory of colors, the Englishman had neither the time nor the inclination to write further. In fact, he wrote to Oldenburg in the cover letter to his second letter to Leibniz, "I hope this will so far satisfy Mr. Leibniz that it will not be necessary for me to write any more about this subject. For having other things in my head, it proves an unwelcome interruption to me to be at this time put upon considering these things."

Indeed, two days after sending the second letter to Oldenburg, Newton wrote to him again, begging, "pray let none of my mathematical papers be printed without my special license." For the next few years, Newton hardly wrote any letters at all, to anybody.

In August 1678, Oldenburg went to Kent for a summer holiday with his wife, and while there they both contracted a severe fever and died. When Oldenburg died the communication between Leibniz and Newton died with him. The correspondence, slow in starting and marked by difficult interruptions with Leibniz suddenly moving countries in the middle, now ended abruptly.

In the ten years that followed, Newton and Leibniz completely lost track of each other. Newton shrunk back into his office in Cambridge University, and Leibniz became mired in the dealings of the court of Hanover—a position he would hold for the rest of his life.

6

The Beginning of the Sublime Geometry

If it takes two to make a quarrel, it takes two men of genius to make a famous quarrel.

—A. R. Hall, *Philosophers at War*

Hanover Germany is a new city today—literally. Destroyed by allied bombing raids during World War II, it was rebuilt from the streets up. It is now home to a large university and a population of about a half million.

A large sign at the airport greets travelers with the salutation WELCOME TO HANOVER, THE CITY OF INTERNATIONAL FAIRS. When asked what type of fairs these were, one local resident said that they were industrial in nature, featuring computers and machines that build machines. Apparently, the fairs were from Leipzig, which had traditionally hosted them until after World War II, when Leipzig wound up in East Germany.

Like the fairs, Leibniz came to Hanover as an industrious transplant from Leipzig. He spent forty years there, the better part of his life, in the service of the dukes of Hanover, engaged in such endeavors as establishing the court library, researching the genealogy of the family, and writing its history.

The house where Leibniz took up residence in 1698 and lived on and off for the last two decades of his life, was built in 1499. Like the town itself, the house was completely rebuilt post–World War II, after it was utterly destroyed in a bombing raid in 1943. Following the war, there were discussions about what to do about the town in general and Leibnizhaus in particular. By the time the decision was made to rebuild the house, a shopping mall and a parking garage had already been built on top of the original site. So instead, construction was undertaken at an alternative site, the present-day Leibnizhaus. The new building had another problem in that the newly chosen site butted up against another building such that, if they had built an exact replica of the house, it would have overlapped its neighbor. Finally, the decision was made to construct a modern building covered with a genuinely old facade.

The new Leibnizhaus opened in the 1980s as part of the University of Hanover, and has a guesthouse, a conference center, and a small museum on the ground floor. At the museum are some original pieces, as well as a painting and a bust of Leibniz, and a casting of his skull.

The building's incredibly ornamental baroque facade dates from 1651, just a few years before Leibniz moved to Hanover. The right side of the facade stands out from the rest of the face of the building, and forms a sort of three-story bay window, accentuated on each level by four fancy columns and decorated up and down with lots of angelic figures. The top of the building has several stepped, orthogonal levels and a series of small windows. The front rooms of Leibnizhaus look out onto a sort of small square. There is a fancy wrought iron monument in front of the building, which is situated on a street crowded with high-end secondhand shops, clothing stores, small restaurants, and a coffee shop or two. The building is flanked by a Spanish restaurant and an antique shop.

All the windows in the front of the building are divided into smaller, crosshatched squares. From the outside of the building, they are barely noticeable, as the building itself is so remarkable in other ways. But from the inside, the square panes dominate the view, which is framed terrifically by the tall windows. The rectangular frames appear dark against the buildings that are visible outside, across the street, echoing the shapes of the building that are themselves very rectangular and Bavarian, with exposed timbers and lots of windows.

Though the facade maintains period accuracy, the inside of Leibnizhaus is very different from the interiors of the mathematician's time. And that is not all that has changed.

Hanover is a college town, and outside is every sign that the university is nearby. The university was, when I visited, the largest in lower Saxony, with more than 24,000 students—not a bad place to visit or to live. But the town was much smaller when Leibniz resided there.

His move into the court of Hanover was not an uncommon path for a man in his position to take. In his day, many people who were smart and social climbers would seek patronage in the courts of Europe. The best way to do this, of course, was to provide ways for the princes and dukes to increase their revenue. Wars, famines, and the lavish courtly lifestyles of the times all took their toll on the noble pocketbook, and a creative thinker who could come up with new schemes for making money was very valuable indeed. Still, probably few were as creative at schemes as Leibniz was.

The irony of Leibniz's life is that, while he might seem so very academic today, he chose not to follow an academic path. Once he finally, reluctantly, arrived in Hanover, he stayed there most of his life, also serving in various capacities in the nearby courts of Celle, Wofenbüttel, Berlin, and Vienna—a career Bertrand Russell once called a lamentable waste of time. But to Leibniz, this pathway really made sense. Despite the fact that he was criticized in the eighteenth century for believing that this was the best of all possible worlds, he spent a considerable amount of time in the seventeenth century hoping to improve it.

Knowing the reality of his day was that power was concentrated in the hands of the few, Leibniz also held the neo-utopian beliefs that those holding this power should be wise and pious men, benevolent leaders who would be best suited to raise mankind to its greatest potential. While it would have been too much to expect that all nobles and hereditary rulers could themselves be wise men, he thought that any change to society should happen within the context of existing political power structures, and wanted to work within these structures. He desired to enlighten the princes, dukes, and other rulers of his day so that they could make the right choices. He was attracted to the job at Hanover because the duke appeared to Leibniz to be wise as well as powerful.

Johann Friedrich's grandfather had been the great Duke William of Lüneberg, also known as William the Pious, who governed with religious discipline and left fifteen children to sort out the spoils of his kingdom among themselves. William the Pious went mad and blind and, on his deathbed, his children drew lots to decide the fate of the ducal lands. William's sixth son, George, won. But George was not fit for the sedate and pious country life his father had established, and he went on a grand tour of Europe, indulging his every whim. Just after he tired of this and returned home, the Thirty Years' War broke out, and George fought with the Holy Roman Empire in Lower Saxony and Italy. When he was finished, he took an abbey at Heldesheim as his personal booty. There he rested, and there he died. His oldest son, Christian Louis, succeeded him and became the new duke, but Christian Louis died childless a few years later. The territory was divided among the three remaining brothers, one of whom was Johann Friedrich. Thus Johann Friedrich became duke of the territory that included Hanover.

At Hanover, Leibniz took command of a library that contained 3,310 books and dozens of manuscripts. However, he was not satisfied with it and proposed a plan to the duke to expand its holdings.

Having just come from one of the most learned centers of Europe, Leibniz was in a good position to claim the breadth of knowledge to be able to do so. In the years to come, he would add thousands and thousands of works to the collection.

For instance, he went to Hamburg in 1678 to look over the library of Martin Fogel. The availability of the Fogel collection was a tremendous opportunity for a book lover like Leibniz and a library builder like the duke, since it had 3,600 rare tomes on natural science and other subjects, so Leibniz convinced the duke to buy it. While he was there, he met Heinrich Brand, who discovered a way to manufacture phosphorus by accident, apparently, when he was following the instructions in an ancient alchemy book for extracting a chemical from urine that could turn silver into gold.

Leibniz convinced the duke to pay Brand to come to Hanover and set up a laboratory to manufacture phosphorus. The key starting ingredient in this process was urine. To produce a substantial amount of phosphorus, Brand needed a grand supply of his starting material. So he had barrels brought into the camps of the region's soldiers, and these fighting men of war supplied him with his precious liquid, which was then shipped to Brand's laboratory. I get this picture when I think about it: German soldiers from a mostly forgotten time standing around, foul mouthed, cackling and filling up the barrel. Liquid gold.

More books was not all Leibniz requested. Within a few months of arriving, he asked for and was given the honor of a promotion to a higher rank of counselor, with an increase in salary. In the beginning, Leibniz was happy enough with his new life to write to some of his acquaintances abroad that he was pleased to be working for the duke who, in addition to being smart and discerning, was wise enough to allow him the freedom to pursue his own endeavors throughout the ticking hours of the day; he gave Leibniz ample time to devote to his intellectual pursuits. Leibniz even wrote one man in Leipzig, a Martin Geier, that he would rather work for Duke Johann Friedrich than enjoy every kind of freedom.

Meanwhile, the duke appears to have been impressed by the

words of the philosopher Antoine Arnaud, a renowned scholar his
new privy counselor knew in Paris, who paid Leibniz the great compliment of saying that the only thing that was possibly holding Leibniz back was his Protestantism.

Still, Hanover was certainly not the throbbing heart of the scientific revolution. Even though it was a large city by German standards,
its population was only around 10,000—as opposed to cities like
Madrid or Amsterdam, which had well over 100,000, or London,
which had somewhere close to half a million. And, despite the fact
that the court at Hanover is described as one of the most elegant and
cultivated in all of seventeenth-century Germany, Leibniz was no
longer in Paris. In Hanover, there was no scientific society compa-
rable to those in London or Paris, and no community of intellectual
peers—except perhaps the duke.

Their tastes happily coincided, and Johann Friedrich is said to have
often joined Leibniz in his physical and chemical studies. Leibniz was
teeming with ideas, and in the duke he found a patron who seemed
willing to sponsor his ideas and had sufficient intelligence himself to
grasp the vision. Together, the two could have been the dream team
of intelligent governance.

Leibniz's grand vision was to bring about improvements to a
universal Christian society through the application of science and
technology. He wrote three memoranda to the duke in 1678, proposing ways to improve everything from agriculture to public administration. He called for an economic survey to gauge the state of the
state in terms of the number of workers and the amount of natural
resources that would serve as the raw data for an analysis for improving economic output; the establishment of an academy to teach
young people commerce; and the creation of something resembling
the modern department store, where common goods could be purchased cheaply in one central place. He recommended that the state
archives be organized under one director—himself, of course—so
that information could be more easily accessed. He called for the creation of a bureau of information that would produce a magazine and

would provide a valuable eBay-like source for people looking to acquire rare goods and services. And he recommended incentives for farmers who followed good farming practices.

The above proposals were followed shortly by one for writing a book to be called *Demonstrationes Catholicae*, which would justify the reconciliation of Catholics and Protestants. At that time, Christianity was fragmented, after more than one hundred uncomfortable years of Protestant reformation that had started with Martin Luther's questioning of papal authority in 1517 and continued when the French preacher John Calvin moved to Geneva in 1536. By the mid-seventeenth century, the influence of Luther and Calvin had spread rapidly throughout Europe, opening up pockets in England, throughout Scotland, in France, the Netherlands, large parts of the Holy Roman Empire, a few parts of Poland and other Eastern lands, and even large settlements in the New World.

Leibniz was not, by any means, the only figure in those days to see the value of reunifying the Christian churches, nor was he filled with unreasonable expectations as to its prospect for success. Nevertheless, he proposed finding some common ground and agreement between the theological systems, mainstream elements of both traditions, and engaged an extensive correspondence with various Catholics and Protestants in this regard.

Leibniz was a chief negotiator in the last quarter of the seventeenth century, to reunify the Lutheran and Roman Catholic. The main obstacle to reunification was that it required the reconciliation of beliefs and practices no longer compatible with one another. These were not necessarily obscure matters of theological philosophy but contentions so basic as to seem absurd. The Catholics, for instance, had to accept that the Protestants should no longer officially be regarded as sinners, and the Protestants had to agree to no longer call the pope the antichrist. (One wonders whether "Your Holiness, the antichrist" would have sufficed.) Not surprisingly, Leibniz found the positions of some of the religious authorities unyielding, and these negotiations, which began in Hanover in 1683, ultimately failed.

Long before his grand unification plans petered out, Leibniz suffered a personal and professional tragedy when Johann Friedrich died in 1679. Leibniz was struck with such sadness that he wrote three different eulogies dedicated to the memory of greatness of his friend and boss—including one in Latin and one in French verse.

Leibniz was confirmed in his position as counselor by the new Duke of Hanover, Ernst August, Johann Friedrich's brother, and immediately began pressing his innovations upon his new employer. He had to tailor these proposals carefully. The new duke was not the philosopher his brother had been. Ernst August was a warrior who was recognized for his bravery. The library languished under the new administration. Ernst August spent a fraction of the amount his brother had on new acquisitions, and most of the money he did spend went toward paying bills left over from purchases predating his accession. Less pious and more rowdy than his late sibling, Ernst August is said to have loved the bottle, his stomach, and women—not necessarily in that order. He was given to long drinking bouts and outlandishness, and, in his youth, he had indulged in all manner of vices in Italy and France.

Ernst August's primary concern was to enhance the power of his position and enrich his already extravagant lifestyle. Money was the fuel that could drive this desire, and Leibniz, recognizing this, responded in the only appropriate fashion—by sending the duke proposals that would increase the revenue stream of the court. Thus, money was the motivation for an ambitious project to drain water from the silver mines in the nearby Harz Mountains.

These mountains had been mined for centuries, and the sites were deep and prone to filling with seeping water. Draining them was a necessary step for continuous mining operations as, during the dry months of the year, rivers and streams dried up and pumps that operated on water power couldn't be powered effectively, severely curtailing production in these dry months. A Dutch mining engineer, Peter Hartzingk, had come up with the idea of draining the mines using a combination of water and wind to keep the pumps operating continuously. In his ingenious design, wind power would be used to raise the water to an underground reservoir that could be opened up and emptied into a lower underground reservoir when the wind was not strong enough to operate the pumps.

Leibniz scoffed at this idea and claimed that he could switch the entire operation of the pumps to wind power alone, and he set about designing and implementing improved and more efficient windmills. If he could employ the wind to pump the water out in a steady outward flow, then the mines could be worked even in the winter months, and the silver could continue unabated to the royal coffers in a steady inward flow.

The increased profit, he suggested, could be also used to fund another idea that he had—the granddaddy of all proposals. He wanted to form an imperial scientific academy so impressive that it would surpass even the Académie des Sciences in France and the Royal Society in London. The academy, which was to be made up of forty-nine other scholars and himself, would then become the greatest in the world. Together the scholars would construct an encyclopedia of all human knowledge, wherein concepts would be collected, analyzed, and reduced to their component pieces, and the ways in which they were combined noted, and finally these same pieces and combinations used to build more concepts. Just as words are made up of letters strung together in a written language—or of a string of sounds in a spoken one—so, too, could ideas be thought of as having been formed by letters of the universal characteristic, or so thought Leibniz. The letters he envisioned were something like the unbreakable atoms of the molecule, the pure ingredients of a sauce, the indivisible organs of the body.

Moreover, the letters were only the beginning. Just as a language has a grammar to the way words are gathered together into sentences, so the ideas constructed with the universal characters obey a grammar. Leibniz and his helpers had only to discover these ideal grammatical rules, and they would be able to resolve all questions, from the greatest to the least, by properly resolving the question into the appropriate symbolic characters and then combining the characters into the logical form their internal grammar dictated. It was to be an analysis of human thought worthy of being thought of as a tribute to human analysis.

The universal language was a bold and beautiful idea, but it would not be an easy feat. Nor would it be without great expense, as Leibniz believed that the learned men of the academy, who no doubt would have been from scattered lands throughout Germany, should be freed of financial concerns by supplying them with stipends and the tools and facilities to conduct their research. That kind of funding would be hard to raise since Hanover, like all the German courts, did not have the advantage of large centralized states with extensive tax bases like France. Despite how much the court at Hanover longed for the greatness of the palace at Versailles, how could they possibly compete? The solution, according to Leibniz, was to increase the production of the nearby mines and pour the windfall into his project.

But first, he needed to drain the mines. His memoranda to the duke were vague at first, merely mentioning that he could increase production without mentioning how, but eventually he disclosed that he would design new pumps to eliminate friction and make the conversion of power more efficient using compressed air. He promised to build new and improved windmills that would work better in a slight wind than the existing ones would in strong gale, by implementing folding sails on the windmills that would open and close to adjust to the adjusting strength of the wind. He also came up with a scheme for a horizontal windmill—something that looks like a waterwheel turned on its side.

When he first proposed these ideas shortly before the old duke died, Johann Friedrich had not been an enthusiastic supporter of them, but he was an enthusiastic supporter of Leibniz, so he had agreed to the Harz project in October 1679, and even had a contract drawn up. When Johann Friedrich died, the project had enough momentum that it continued under the new duke, who was all too happy to back the venture financially—at least at first. Even so, he made Leibniz assume some of the costs of building the windmill.

Leibniz was to continually face bad cost overruns and unanticipated expenses. His original estimate of 330 taler had ballooned by the middle of 1683 to a cost of 2,270 taler. And from its inception the project was plagued by infighting. The mining office opposed Leibniz every step of the way. Probably because of their opposition, he began to suspect that his efforts were being sabotaged. He complained to Ernst August that the officials were putting up roadblocks at every juncture and poisoning the workers against him by using lies and threats. The mining office, for their part, poured an equal amount of scorn on Leibniz in their reports to the duke.

Ernst August grew tired of the project after the costs had ballooned and the project had failed to produce results by 1683, and, at the end of that year, cut off his funding for the project. Thereafter, Leibniz had to continue on his own dime. Leibniz did a series of tests in 1683, 1684, and again in 1685 with only partial success. Machines constantly broke down and caused extensive delays, requiring costly repairs. The fickle wind blew and ceased and made even testing the system an ordeal. By the middle of 1684, the weekly report of the mining office was filled with nothing but complaints about the project, and Leibniz faced what he perceived to be a worker's revolt. He blamed the failures on the workers and administrators at the mines, whom he suspected feared for their livelihoods and sabotaged the project and, with it, progress.

Finally, on April 14, 1685, the duke pulled the plug entirely and ordered Leibniz to end construction of his windmills immediately and forever.

Whatever the cause of the project's failures—the heavy expenditures; the initial or eventual lack of support by everyone else concerned, not to mention the uncooperative weather—it also brought about some unanticipated successes. It inspired Leibniz to visit many mining operations on his extensive travels around Europe. He had thrown himself into the work, studying and composing a review of all aspects of mines—from their management to the chemistry of the processes to the geology of the lands. Whenever he went to a region, he tried to make time in his schedule to visit a mine, and he became an expert in mining operations. He even came up with a scheme for altering the bar composition. The silver from Hanover's mines was superior, Leibniz asserted, and so it should be mixed with an appropriate amount of some other ore when it was melted and cast into bullion.

Moreover, in the course of his investigations, Leibniz became interested in the rocks and how they got to be there. It has been said that during his subsequent travels, he never missed an opportunity to study fossils and geological formations. Leibniz looked at the minerals for evidence of their origins, and his insights were at times astounding. When he found an enormous prehistoric tooth in 1692, for instance, he took it as proof not of some ancient monster, but rather as evidence suggesting that oceans once covered the earth. He also proposed the theory that the early earth was molten. In some ways, Leibniz was the father of geology, because he wrote one of the first physical descriptions of the earth, anticipating modern earth science.

Despite his expertise and enthusiasm, his windmill project was an abysmal failure—it failed in its primary goal of drawing water out, producing extra revenue, and enabling the funding of the forty-nine scholars. It was a bit of a financial bomb for Leibniz as well. He spent a small fortune on the Harz project.

MEANWHILE, NEWTON HAD crawled into a deep hole of his own. He was moving steadily away from science and mathematics and into theological and alchemical endeavors, which had consumed him for most of the late 1670s and early 1680s. As much as he was repelled by the controversies surrounding his optical experiments, he was drawn toward these other subjects, which he regarded as highly important and that would variously occupy and consume him for the rest of his life.

He spent a great deal of energy in his alchemical research—untold hours collecting and copying alchemical texts, and working on an extensive chemical index. Manually databasing hundreds of topics, each with references to more than one hundred alchemical texts, plus other commentary, this was an exercise in tediousness. Reading these texts today is nearly impossible. Some of the writings are bizarre, especially to a layperson—full of so many strange symbols and references to mythology that one might have though Newton mad. In fact, these symbols were annotations to denote different elements or substances to be combined, such as lead, copper, or mercury.

Newton was equally drawn to matters of theology. He wrote interpretations of biblical revelations, and worked for years on such projects as elucidating the prophesies of Daniel and John. He was convinced, for instance, that the scriptures had become corrupted during the fourth and fifth centuries. He wrote a few treatises on the subject of the trinity, such as one that he wrote to a friend in 1690 in which he explained, "Since the discourses of some late writers have raised in you a curiosity of knowing the truth of that text of Scripture concerning the testimony of the three in heaven . . . I have here sent you an account of what the reading has been in all ages, and by what steps it has been changed, so far as I can hitherto determine by records. And I have done it the more freely because to you who understand the many abuses which they of the [Catholic] Church have put upon the world. . . . "

He was something of a historian, and set out to correct ancient chronology and to improve it by basing it on mathematical principles. Newton was driven to matching historical facts with biblical references and to elucidating the details of history in general. He concluded, for instance, that the date given for the fall of Troy (then determined to be 1184 BCE) was wrong. He dated it as 904 BCE Newton is also said to have been perhaps the most knowledgeable authority ever on the barbarian invasions in the fifth and sixth centuries. He studied writings from a variety of traditions extensively in order to reproduce the plans of the temple in Jerusalem; concerned with determining its exact dimensions, he examined ancient texts in which the temple was described and translated the ancient measurements into modern lengths.

When he died, Newton's chronology work was recognized to be some of his most important, so much so that an unauthorized version of this historical research was published in 1725 in France by Nicolas Fréret. The official edition of the chronology came out a few years later, in 1728, just after Newton died. It was a history of mankind from the time of Alexander the Great, including Greek, Assyrian, Egyptian, Babylonian, and Persian chronologies, which makes it sound deceptively interesting.

These alternative studies together help to round out the figure of Newton. like many great historical figures, Newton is an enigma. Not because he kept his work private from his wife or worked secretly for some government's war efforts. He never married, in fact, and his political world revolved around scientific intrigue more than it did around the wars and problems of his day. Newton was an enigma because he contributed so much to humanity through his science ad yet spent so many years in endless contemplation of religious and alchemical pursuits. Even though these endeavors really fit naturally with the time he was alive, it seems strange that such a brilliant scientist would have wasted so much time on alchemy, theology, and his chronology of historical and biblical events!

During the 1680s, while Leibniz was seemingly consumed by his windmill project, Newton's calculus work was gathering layer after layer of thick dust. But—Leibniz had not spent every moment in the mines. He was about to publish the first paper ever in the field of calculus and thus fire the first shot in the calculus wars.

———————

MATHEMATICS, FOR LEIBNIZ, had the power of demonstration. In the early 1690s, Prince Gasto of Florence, whom Leibniz had met during his travels through Italy, had sent him a problem for constructing a certain geometrical shape that he needed to solve, and the German was able to come up with a solution in just a few hours. But Leibniz dreamed of a mathematics that reached much further than the subject we today think of as math (as a pure discipline on its own, or one that finds application mostly in scientific applications, the social sciences, and so on). Leibniz saw possibilities for mathematics in ways that can hardly be imagined.

He thought that it might be possible to create an aesthetic calculus that would allow artists to create great works of art the way that a person can solve an equation by plugging in numbers and calculating. He even thought the same general approach could be used for creating poetry and music, which he defined as "an arithmetic of the soul, which knows not that it reckons." That he never went anywhere with any of these other calculi in no way detracts from what he did with calculus, introducing it to the world before anyone else.

The story of his publication started during the Harz mines project, when Leibniz played host to Otto Mencke, a professor he knew from Leipzig where he had grown up. Mencke had an idea to start a scholarly journal that would keep the intellectuals in Germany abreast of the latest discoveries in the German states and throughout Europe, and Leibniz was a big supporter of this idea. He became a cofounder of this journal with Mencke, and in 1682, the *Acta Eruditorum Lipsienium*, or "The Acts of the Scholars of Leipzig" or sometimes as "Transactions of the Learned" began publication as a monthly scholarly journal.

It was the first scientific journal in Germany, and Leibniz was closely associated with it, publishing in it all the way up until his death in 1716. This was an important thing for Leibniz, who had experienced some difficulties in publishing and had tried repeatedly over the course of three years, from 1677 to 1680, to have one of his mathematical treatises published in Paris or Amsterdam without success. But now he could publish freely in this new organ, and he often contributed papers to it—including many of the key documents in the calculus wars.

Even in the early 1680s—years in which Leibniz witnessed the unhappy unraveling of his mine shaft windmill idea—he was so prolific that he might publish an important paper in mathematics one month and a seminal paper on his philosophy the next. In October of 1684, right in the most troubling time of the mining project, he published a paper whose short title is *Nova Methodus Pro Maximis et Minimis* (New method for maxima and minima) in the *Acta Eruditorum*. This was the first calculus publication anywhere in the world, and in it, Leibniz gave the rules for differentiation.

In the cover letter for the paper to his friend Mencke, he wrote that his calculus "will be of the utmost use in the whole of mathematics." One of Leibniz's later admirers gushed over the publication, "[In] 1684 he proceeded to publish the results of his labors in the *Acta Eruditorum*; and thereby called forth the admiration of the whole scientific world at the richness and brilliancy of his discovery."

In actuality, the paper was more complicated. It was modeled after a half-century old work by Descartes called *Geometry*, and stylistically it was difficult to read. Jacob Bernoulli called it an enigma rather than an explanation. Though it was a mere six pages long, its full title was worthy of a much longer piece: "A New Method for maxima and minima as well as Tangents, Which Is Neither Impeded by fractional nor irrational quantities, and a Remarkable Type of calculus for them," as translated into English.

But it had treasures aplenty. In the paper, Leibniz performed feats of mathematics with ease, such as deriving Snell's law of sines. "Other very learned men," he wrote boldly, "have sought in many devious ways what someone versed in this calculus can accomplish in these lines as by magic." He solved with ease a problem that Descartes was unable to solve in his lifetime. "And," Leibniz continued in the paper, "this is only the beginning of much more sublime geometry, pertaining to even the most difficult and most beautiful problems of applied mathematics, which without our differential calculus or something similar no one could attack with any such ease."

Significantly, Leibniz had no historical introduction to his paper.

Had he had one, he might have mentioned the work that developed his methods and the communication he carried out with Newton nearly a decade earlier. In the paper, Leibniz made no reference at all to the correspondence, and nowhere does he give Newton credit in this or any subsequent publication, and this may have been a mistake. Had he acknowledged Newton in some way, Newton may not have come back at him years later. But he had no such language. Instead he just plunges into a terse explanation of his own methods without ever once mentioning Newton.

Though Leibniz did not mention Newton in the article, he did mention him in the cover letter he sent to his friend Mencke in July of 1684. "As far as Mr. Newton is concerned, I have his and the late Mr. Oldenburg's letters, in which they do not dispute my quadrature with me, but concede it," Leibniz wrote. "I do not believe, either, that Mr. Newton will claim it for himself, but only some inventions relating to infinite series which he has in part also applied to the circle." These inventions, Leibniz tells his friend, were first discovered by Mercator, then developed by Newton, and then continued by Leibniz "by another way."

In this cover letter, Leibniz anticipated the calculus wars and dismissed them at the same time, determining that he had come up with one method, and Newton another. "I acknowledge," he wrote, "that Mr. Newton already had the principles from which he could well have derived the quadrature, but all the consequences are not come upon at once: one man makes one combination and another man another."

Leibniz could not be blamed in a sense if he underestimated Newton, since for Leibniz, the second letter Newton sent in 1676 held little more than a "bare enunciation" of concepts, none of which were even new to him. Nevertheless he seems to have recognized that Newton was in possession of certain mathematical techniques parallel to his own calculus, even if he was never quite satisfied in his desire to find out what Newton's method of fluxions was exactly. When the calculus wars were in full tilt, in what would be the most bitter irony for Leibniz, Newton would turn things around and claim that he was in fact so explicit in explaining his fluxions to Leibniz that it was essentially what allowed Leibniz to put together his calculus.

This would not be for years, of course. In 1684, when Leibniz published his calculus, Newton had more or less abandoned mathematics. But he was about to be pulled back into it in a big way. Meetings and exchanges were taking place that would lead Newton to publishing the book for which he is most famous, the "Mathematical Principles of Natural Philosophy," or *Principia*.

The earliest of these exchanges took place in the late 1670s and were initiated by Hooke, who extended Newton an olive branch in a letter he wrote on November 24, 1679. "I hope therefore that you will please to continue your former favors to the Society by communicating what shall occur to you that is philosophical, and in return I shall be sure to acquaint you with what we shall Receive considerable from other parts or find out new here," Hooke wrote to Newton.

In the same letter, he tried to make amends for their earlier troubles. "I am not ignorant that both heretofore and not long since also there have been some who have endeavored to misrepresent me to you and possibly they or others have not been wanting to do the like to me, but difference in opinion if such there be (especially in philosophical matters where interest hath little concern) me thinks should not be the occasion of Enmity—tis not with me I am sure," Hooke wrote. "For my own part I shall take it as a great favor if you shall please to communicate by Letter your objections against any hypothesis or opinion of mine." And then he added, "particularly if you will let me know your thoughts of that of compounding the celestial motions of the planets of a direct motion by the tangent & an attractive motion towards the central body."

This last bit was really the reason why Hooke was interested in chatting Newton up. He knew that Newton was an outstanding mathematician and natural philosopher, and Hooke had become interested in a subject about which he suspected Newton had a great deal of insight—the gravitational nature of planetary motion. Hooke

wrote again on January 17, 1680 reiterating his interest in the properties of the path a body would take under the influence of a central attractive power—essentially what path would something like a
comet or the Earth follow in its course around the sun if it were
attracted to the gravitational pull of the sun. "I doubt not but that by
your excellent method you will easily find out what that Curve must
be, and its properties, and suggest a physical reason of this proportion," Hooke wrote. "If you have had any time to consider of this
matter, a word or two of your thoughts of it will be very grateful to
the society (where it has been debated)."

Hooke may have found it easier to suggest that the Royal Society at large was interested in Newton's opinions than admitting that
he was the primary one who was. As secretary of the Royal Society, he certainly had the authority to speak for the body as a whole.
And the exchanges were, on the surface, very cordial, with Newton
signing his letters, "Your very much obliged & Humble Servant Isaac
Newton" and Hooke signing, "Your most affectionate humble Servant Robert Hooke." But they never went anywhere—that is, until
a few years later when Edmond Halley came into the picture.

Halley met Hooke and Christopher Wren in a coffee shop some
time in the spring of 1684. Coffeehouses flourished in London in the
seventeenth century, and by the end of the century there were thousands. These provided a forum for meetings, and I imagine them to
be like the best coffee shops today with strange men reeking of
tobacco meeting each other and sitting leaning over broad thick
tables stained black with coffee bean oil. Halley was curious about
the comet that today carries his name, and he posed a simple question to these two other men: what sort of path would a celestial
object like a comet take?

Hooke had a physical explanation, and it was the right one.
Celestial objects, he said, would follow an inverse square law of attraction. Wren, perhaps unconvinced, asked Hooke to demonstrate how
he knew it was an inverse square law, but Hooke demurred. Wren
challenged Hooke to prove it, promising that he would reward him

with a valuable book worth forty shillings if he could do it, but Hooke had no such mathematical proof, so this was not a bet that he could accept. Instead he declined. Meanwhile, Halley sat there unsatisfied. He thought he had the answer, but how could he be sure?

Wren told Halley that a certain mathematics professor he knew in Cambridge named Isaac Newton might be able to answer the question, and so several months later, in August, just as the printers in Germany were about to press Leibniz's famous first calculus paper to ink, Halley took a trip to Cambridge.

The dusty, bumpy, meandering fifty-mile ride must have been hell for Halley inside a shaking deathtrap of a seventeenth-century horse-drawn carriage. Today, it's a breeze. For a pocketful of small bills, anyone can buy a ticket for a train ride to Cambridge from London that leaves two to three times an hour, and takes about forty minutes non-stop. The scenery on the journey is probably in many ways still similar to how it was in Newton's day. You pass through rolling fields and farmlands separated by old stone walls. Aluminum siding, diesel tractors, and the occasional satellite dish are the only objects that give the landscape a more modern appearance.

When Halley got to Cambridge and walked through the grand gate at Trinity College, he sought out Newton and asked him the same question he had posed to Wren and Hooke months before: What sort of path does a celestial body follow? Newton answered immediately: an ellipse. The orbit of the planets around the sun follow an inverse square law, and the path is elliptical. It was a simple answer that would change both men's lives forever.

Halley was "struck with joy and amazement" at hearing Newton's words. This was the sweetest music to Halley's ears to hear Newton say what Hooke had already told him. But could Newton prove it? Halley asked him how he knew. Newton replied that he knew because he had done the math. He had calculated it.

At this, Halley immediately asked to see the publications. Newton had worked many of these things out years before and was not exactly sure where to find the calculations—certainly not while Halley was waiting anxiously in his rooms. So he bid Halley return to London and promised to send the calculations afterward. This was a promise that Newton kept, sending two proofs after Halley left. He also wrote a short book, *De Motu Corporum* (On the Movements of Bodies), which he sent to Halley as well. Halley, recognizing its importance, began cajoling Newton to write more.

Newton did. He started in 1685 and sent the first part to the Royal Society in time to be recorded in the April 28 minutes of the Royal Society. While some may think that Halley's greatest contribution was predicting the return of the comet he ultimately gave his name to, one could argue that in fact his greatest accomplishment was to convince Newton to publish one of the greatest books ever written—the *Principia*.

In fact, Halley did not only cajole Newton into writing the *Principia*, he also oversaw the production of the book and personally underwrote the expense of publishing it in 1687, since the Royal Society could not scrape together the funds to do so. "I have at length brought your Book to an end," Halley wrote to Newton on July 5, 1687, "and hope it will please you." Halley gave Newton twenty-seven copies and provided forty for booksellers in Cambridge, which could be had, back then, for a few shillings.

And Halley wrote proudly to King James II, in July of 1687, "And I may be bold to say, that if ever Book was so worthy of the favorable acceptance of a Prince, this, wherein so many and so great discoveries concerning the constitution of the visible world are made out, and put past dispute, must needs be grateful to your majesty; being especially the labors of a worthy subject of your own, and a member of that Royal Society founded by your late royal brother for the advancement of natural knowledge, and which now flourishes under your majesty's most gracious protection."

While Newton was first starting to work on the *Principia*, Leibniz's work was spreading in Europe and had reached across the English Channel. A Scotsman, John Craig, who lived in Cambridge and was a friend of Newton's, published the first publication on calculus to appear in England in 1685, a year after Leibniz had published his paper. Craig wrote a book, *The Method of Determining the Quadratures of Figures*, which described Leibniz's work on differentials and used Leibniz's notation. This effectively introduced England to calculus—or at least most of England, since Newton had been sitting on his own methods for two decades.

Craig was a mathematical enthusiast and something of a forgotten player in the invention of calculus. He published more on the subject than perhaps anyone else alive in his time. In addition to his 1685 book, he wrote another in 1693, and he also contributed articles on calculus to the *Philosophical Transactions of the Royal Society* in 1701, 1703, 1704, and 1708. Perhaps because he was so heavily indebted to both Newton and Leibniz, his is not a name readily associated with calculus today.

Nor is he a central fighter in the calculus wars—probably because he was willing to seek out and acknowledge his sources of inspiration. Craig had spoken with Newton prior to publishing and had gotten the binomial theorem from him prior to the 1685 book. In his 1693 book, Craig wrote what can be regarded as the model of elegant acknowledgment of Leibniz. "In order not to seem to assign too much to myself or detract from others, Craig wrote, "I freely acknowledge that the differential calculus of Leibniz has given me so much assistance in discovering these things that without it I could hardly have pursued the subject with the facility I desired."

Leibniz, aware of Craig's 1685 book and of the efforts of mathematicians and natural philosophers elsewhere in Europe, was inspired to send his second paper on calculus in 1686 to the *Acta* with the title "On Recondite Geometry and the Analysis of Indivisibles and Infinities." It was on what he thought of as the inverse of differentiation—integration. He began the paper by boasting that the methods he presented in his previous paper "won no slight approval from certain learned men and are gradually indeed being introduced into general use." And in this second, longer paper, Leibniz promised to illuminate calculus further.

"Like powers and roots in ordinary calculations, so here sum and differences . . . are each other's converse," Leibniz wrote. Both his 1684 and his 1686 papers are noteworthy as the first published descriptions of calculus and for their introduction of notation for differentiation and integration (the twin tools of calculus-based analysis) that are still in use today—though the word "integral" and integral calculus, now commonly used, was not actually mentioned in the 1686 paper. In fact, Leibniz had never intended to call his "recondite geometry" integral calculus. The term integral was first used in a paper by one of the Bernoulli brothers in 1690 and "integral calculus" first appeared as a term in a paper written by Johann Bernoulli with Leibniz in 1698.

The year 1686 was one in which things really crystallized for Leibniz. He published his famous "Discourse on Metaphysics" that year, his first actual systematic description of his philosophy, and this enabled him to begin a correspondence with Antoine Arnaud—something that he had tried to do nearly twenty years before. Leibniz sent Arnald the title headings from his "Discourse on Metaphysics" as a way of opening up the conversation, and what transpired is one of the most famous discourses in the history of philosophy, the Arnaud–Leibniz correspondence, which is still in print today.

In a way it is strange to think about how this philosophical work inspired a lengthy and interesting discourse because the same thing could have happened with Leibniz's mathematical papers. They might have had the same effect this philosophical paper did and have been the cause for a discourse on mathematics to begin between him and Newton. But they had no such effect.

Newton was busy writing the *Principia*, a mammoth and all-consuming project. In fact, it's fair to say that in these years, Newton had a second, midlife *anni mirabiles* in which he wrote the work in a mere eighteen months.

On May 22, 1686, Halley wrote proudly to Newton, "Your Incomparable treatise . . . was by Dr. Vincent presented to the R. Society on the 28th past, and they were so very sensible of the great honor you do them by your dedication, that they immediately ordered you their most hearty thanks, and that a council should be summoned to consider about the printing thereof."

Around the same time, Halley was the bearer of bad news. When Hooke found out about the *Principia*, he was furious. He had sent Newton a letter some six years before, and he was not about to sit quietly with his suspicion that Newton was stealing his thunder yet again.

"There is one thing more that I ought to inform you of," Halley wrote to Newton, "that Hooke has some pretensions upon the invention of the rule of the decrease of gravity, being reciprocally as the squares of the distances from the center. He says you had the notion from him, though he owns the demonstration of the curves generated thereby to be wholly your own."

Hooke wanted Newton to give him his due credit, and Halley wrote to Newton and politely suggested that he do it. "Mr. Hooke seems to expect you should make some mention of him, in the preface," wrote Halley.

Newton bristled at the notion. After Halley wrote to Newton sending him the first proof of the *Principia*, Newton wrote back on June 20, 1686 asking that his intelligence not be insulted. "I hope I shall not be urged to declare in print that I understood not the obvious mathematical conditions of my own Hypothesis. But grant I received it afterwards from Mr. Hooke," Newton wrote. After pages of defense in the dispute with Hooke, Newton finally addresses Halley's letter by saying, "The Proof you sent me I like very well."

Then he adds a note consisting of several more pages: "Since my writing of this letter I am told by one who had it from another lately present at one of your meetings, how that Mr. Hooke should there make a great stir pretending I had all from him & desiring they would see that he had justice done him. This carriage towards me is very strange & undeserved." Newton was so infuriated that he threatened to kill the third part of the *Principia* altogether. Eventually, he did calm down and capitulated to Halley's suggestion that he mention Hooke, but only in the context of Christopher Wren and Halley.

In the *Principia*, Newton also mentions his earlier exchange of letters with Leibniz. "When, in letters exchanged between myself and that most skilled geometer G.W. Leibniz ten years ago, I indicated that I possessed a method for determining maxima and minima, of drawing tangents and performing similar operations ... that famous person replied that he too had come across a method of this kind, and imparted his method to me, which hardly differed from mine, except in words and notation."

These words would become bandied about by both sides in the calculus wars years later, but in 1687, they went almost without notice. That year, just as a decade before in the 1670s, was a lost moment. Just when Newton may have otherwise taken a hard look at what Leibniz was printing and what people were saying about calculus, he was distracted by more troubles with Hooke. Instead of starting a conversation that could have resulted in acknowledging their co-invention of calculus, they became aware of the publications of the other and began to form, on opposite sides of the English Channel, a quiet competition—quiet for now at least.

7

The Beautiful and the Damned

▪ 1687–1691 ▪

The benefits which, in the course of almost half a century, would have accrued to science from the harmonious connection, thus unceremoniously dissolved, of these two great philosophers, can hardly be too highly estimated.

—John Milton Mackie, from
Godfrey William von Leibniz, 1845

Samuel Pepys lived a charmed life. He was not one of the greatest men of his day and yet his name still rings out today because he was a witness, through his diaries, to one of the most interesting times in the history of England. And the short period of his life when he was the head of the Royal Society was no exception. It was during his brief presidency, slightly more than a year— one of the shortest tenures anyone ever spent in that position, in the 350-year history of the Royal Society—that Newton finished the *Principia*. And because Pepys oversaw the delivery of it, the "Julii 5, 1686" imprimatur on the title page carries the name *S. Pepys Reg. Soc. Praeses*. Along with some information of the printers and, finally, the date of publication MDCLXXXVII (1687).

Many commented swiftly on the importance of the book. David Gregory wrote to Newton on September 2, 1687, that "having seen and read your book I think my self obliged to give you my most hearty thanks for having been at the pains to teach the world that which I never expected any man should have known. For such is the mighty improvement made by you in the geometry, and so unexpectedly successful the application thereof to the physics that you justly deserve the admiration of the best Geometers and Naturalists, in this and all succeeding ages."

What a book it was! Its more than five hundred pages of seventeenth-century scholarly Latin were filled with complicated diagrams, illustrations, tables of astronomical observations, geometrical drawings, and a parade of propositions, problems, corollaries, definitions, and scholia, the *Philosophiae Naturalis Principia Mathematica* ("Mathematical Principles of Natural Philosophy"), or simply *Principia* was a book that was destined to have a major impact on science. It would become one of the single greatest scientific books ever written, containing one of the greatest bodies of knowledge ever conceived: Newtonian or "classical" mechanics, an understanding, with mathematical descriptions, of the mechanics of motion, which is still the gateway subject to a physics degree today.

Anyone who has ever written a technical paper or an original work of science, no matter how broad in scope, can appreciate the sheer magnitude that was the *Principia*. In it, Newton put forward the laws of mechanics as applicable on the earth as in outer space. He laid out proofs of Kepler's laws based on his original work of centers of mass and gravity. He also used gravity to explore the attraction between two massive objects. This allowed him to explain how Jupiter and its moons interact. Newton explained the flattening of the Earth at its poles and the Earth's bulging at its equator. He described phenomena that become the basis for fluid mechanics, and considered such topics as resistance to motion, pendulum motion with and without resistance, and the motion of waves. He worked out the theory of the tides, explaining it in terms of gravity and the moon's pull on the Earth.

Finally, looking to the rest of the solar system, he described the motions of the planets and explained the precession of the equinoxes. He wrestled with comets and showed that they are part of the solar system. He estimated the density of the Earth and calculated the masses of the planets, the sun, and the Earth, and he was actually close on the mass of the earth. He disposed of the vortex theory of planetary orbitals favored by Descartes and many others in the seventeenth century, and went on to reconcile the orbits of Jupiter's and Saturn's moons with his theory of universal gravitation.

Most significantly, he developed his theory of universal gravitation—that objects are attracted to one another by virtue of the force of gravity. This was by far the most radical idea presented in the *Principia*—the notion that there was some force by which every mass in the universe attracts every other mass. It has been called the most important scientific discovery of all time, and because Newton used the theory to conceive of the universe as an expanse of objects interacting with one another through this gravitational force, he has been called the father of modern astronomy—even though he was unlike other astronomers of his day, in the sense of their looking through a telescope and observing the motions of heavenly bodies.

Universal gravitation was truly a revolution in science, flying in the face of both conventional wisdom and logic. To the student of science, this discovery is an inspiring example of a brazen young scientist at his most successful, who through a combination of hard work and genius puts together something that can truly be considered a new paradigm. To the philosopher, universal gravitation is pregnant with profound consequences—every particle in the universe attracting every other particle. And to the historian, the gradual way that resistance gave way to acceptance and then finally to championing of Newton's theory by others in his lifetime and after his death is a remarkable story.

Finally, to the historian of science, the *Principia* represents a turning point. During the seventeenth century, it wasn't just that people began to change their views of the world. They were beginning to realize that they could also change how they arrived at those views in the first place—through empiricism, i.e., observation and data. But while the ability to comprehend the world through reason and to quantify it through observable measurement was not something unique to Newton, it came into full flower with the *Principia* because he kept to observations as a way to formulate the nature of something and describe it mathematically.

To avoid being embroiled in disputes the way that he was over his optics, Newton felt no need to justify gravity in the *Principia*. He later wrote to a man named Bently that he did not pretend to know the cause of gravity. In fact, Newton wrote in the *Principia* the famous phrase, *Non fingo hypotheses*, or "I do not invent hypotheses."With that phrase in mind, he resisted guessing as to the *nature* of gravity and restricted himself, instead, to describing the *behavior* of gravity. He believed in the experimental over the hypothetical, even when the hypothetical made more sense.

It was not always so easy for his contemporaries to accept that a theory that didn't logically make sense. This was made obvious in an extremely favorable review of the book that appeared in a journal on the continent, which pointed out the good things about the *Principia*. "The work of Mr. Newton is the most perfect treatise on mathematics that can be imagined, it not being possible to provide a more precise or more exact demonstration than those he gives," it read. Nevertheless, the review also put forward this criticism: "He was not considered the principles in question as a physicist but purely as a geometer."

Others, most notably Leibniz, were even less enthusiastic about the idea of gravity and went on to reject the notion. Leibniz could never accept the basic premise of Newton's worldview—that outer space is essentially a vacuum and that the Earth and the planets revolve around the sun by virtue of gravitational attraction. For Leibniz, theory based upon observation was not enough. He could not accept how gravity might act upon an object millions of miles away through apparently empty space.

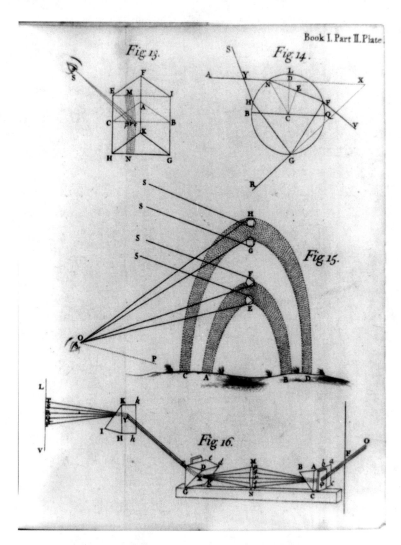

Diagrams of optical phenomena from *Opticks* by Isaac Newton.

SOURCE: Library of Congress.

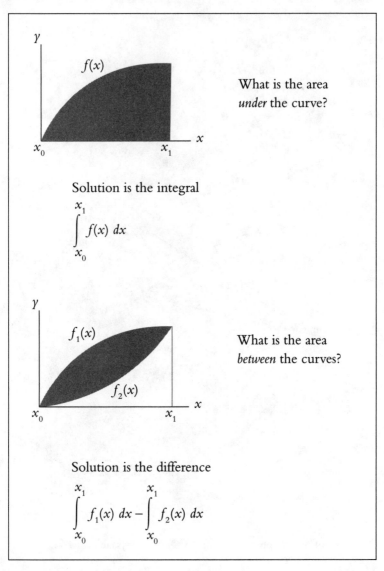

What is the area *under* the curve?

Solution is the integral

$$\int_{x_0}^{x_1} f(x)\ dx$$

What is the area *between* the curves?

Solution is the difference

$$\int_{x_0}^{x_1} f_1(x)\ dx - \int_{x_0}^{x_1} f_2(x)\ dx$$

Tough problems that calculus solves with ease #1.

Isaac Newton.
SOURCE: Royal Society.

Gottfried Wilhelm
Leibniz.
SOURCE: Royal Society.

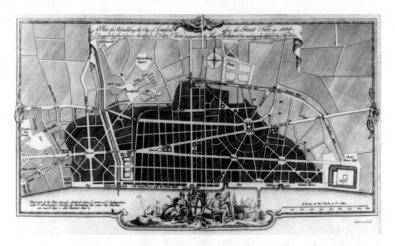

Christopher Wren's plan for rebuilding London after the fire of 1666 was impressive—but so was Isaac Newton's plan for rebuilding the world based on universal gravitation. SOURCE: Library of Congress.

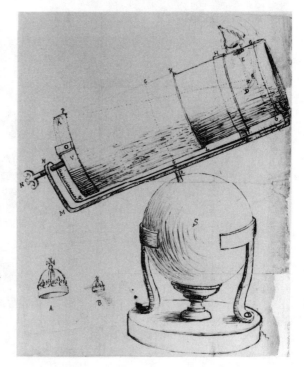

Newton's own drawing of his reflecting telescope.
SOURCE:
Royal Society.

about three foot radius (fuppofe a broad Object-glafs of a three foot Telefcope,) at the diftance of about four or five foot from thence, through which all thofe colours may at once be tranfmitted, and made by its Refraction to convene at a further diftance of about ten or twelve feet. If at that diftance you intercept this light with a fheet of white paper, you will fee the colours converted into whitenefs again by being mingled. But it is requifite, that the *Prifme* and *Lens* be placed fteddy, and that the paper, on which the colours are caft, be moved to and fro ; for, by fuch motion, you will not only find, at what diftance the whitenefs is moft perfect, but alfo fee, how the colours gradually convene, and vanifh into whitenefs, and afterwards having croffed one another in that place where they compound Whitenefs, are again diffipated, and fevered, and in an inverted order retain the fame colours, which they had before they entered the compofition. You may alfo fee, that, if any of the Colours at the *Lens* be intercepted, the Whitenefs will be changed into the other colours. And therefore, that the compofition of whitenefs be perfect, care muft be taken, that none of the colours fall befides the *Lens.*

In the annexed defign of this Experiment, A B C expreffeth the Prifm fet endwife to fight, clofe by the hole F of the window

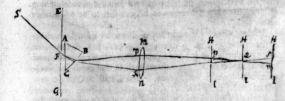

E G. Its vertical Angle A C B may conveniently be about 60 degrees : *M N* defigneth the *Lens.* Its breadth 2½ or 3 inches. S F one of the ftreight lines, in which difform Rays may be conceived to flow fucceffively from the Sun. F P, and F R two of thofe Rays unequally refracted, which the *Lens* makes to converge towards Q, and after decuffation to diverge again. And H I the paper, at divers diftances, on which the colours are projected : which in Q conftitute *Whitenefs,* but are *Red* and *Yellow* in R, r, and ɼ, and *Blew* and *Purple* in P, p, and ϖ.

If

A page from the *Philosophical Transactions of the Royal Society* showing the experiment that led Newton to conclude that white light is made up of rays of different colors.

SOURCE: Library of Congress.

Engraving of a fly,
as seen through the
microscope. From
Hooke's *Micrographia*.

SOURCE: Library of Congress.

Christian Huygens.
SOURCE: Royal Society.

A model of Leibniz's calculating machine.
SOURCE: Gottfried Wilhelm Leibniz Bibliothek, Niedersächsische Landesbibliothek.

Henry Oldenburg.
SOURCE: Royal Society.

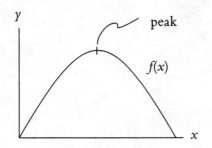

Find the peak
of the curve
$f(x)$

Solution: Take the derivative of $f(x)$,
set equal to zero, solve

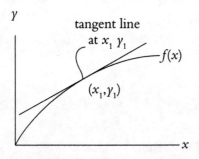

Draw a tangent
to the curve $f(x)$
at any point

Solution: Slope of the tangent line
is equal to the derivative
of $f(x)$ at that point

Tough problems that calculus solves with ease #2.

Leibnizhaus prior to its destruction in WWII, where Leibniz spent his final days. SOURCE: Library of Congress.

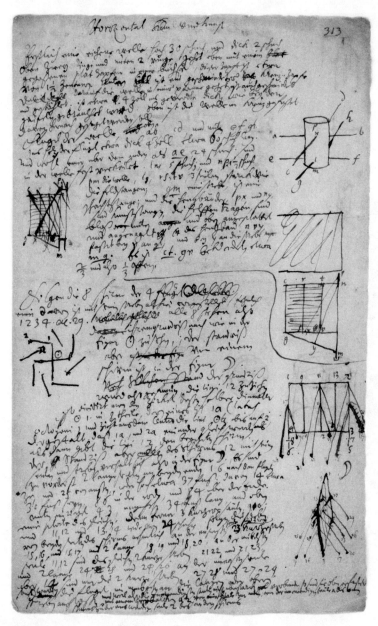

Leibniz's notes on his horizontal windmills. SOURCE: Gottfried Wilhelm
Leibniz Bibliothek, Niedersächsische Landesbibliothek.

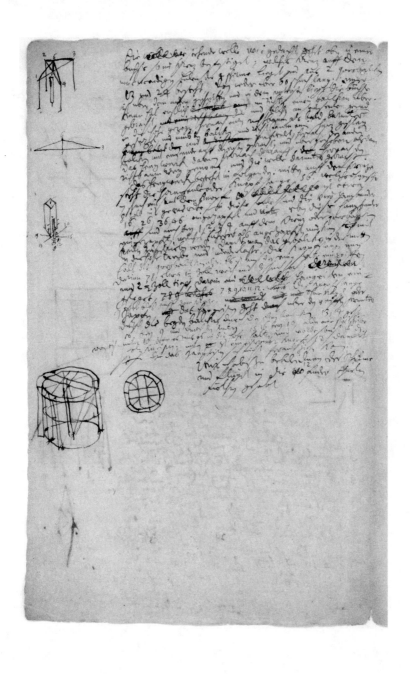

Edmond Halley. SOURCE: Royal Society.

Nicholas Fatio de Duiller. Source: Library in Geneva.

John Wallis. Source: Royal Society.

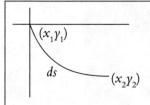

A particle starts from rest and moves under a constant gravitational acceleration from one point $(x_1 y_1)$ to another $(x_2 y_2)$ in the shortest time

Formulating this problem with calculus is easy.

$$time = \frac{distance}{velocity}, \text{ or}$$

$$time = \int_{x_1 y_1}^{x_2 y_2} \frac{ds}{v}$$

where ds is a differential distance along the path.

The Pythagoream theorem has

$$ds^2 = dx^2 + dy^2$$

or

$$ds = \sqrt{dx^2 + dy^2}$$

And the velocity can be determined from the principle of the conservation of Energy

$$\frac{1}{2}mV^2 - mgy = 0$$
$$\text{so } v = \sqrt{2gy}$$

Thus the integral can be written

$$time = \int_{x_1 y_1}^{x_2 y_2} \frac{\sqrt{dx^2 + dy^2}}{\sqrt{2gy}}$$

Solving this integral, on the other hand, is a whole lot harder.

Brachistochrone problem

George Ludwig,
who later became
George I, King of England,
governed Hanover during
Leibniz's final years.
SOURCE: Library of Congress.

When Newton took over the British Mint, it was
housed in this row of buildings in the Tower of
London. PHOTOGRAPHER: Jason S. Bardi.

Part of a letter in Leibniz's own hand describing some of his calculus work. SOURCE: Gottfried Wilhelm Leibniz Bibliothek, Niedersächsische Landesbibliothek.

A copy of the *Charta Volans*.

SOURCE: Gottfried Wilhelm
Leibniz Bibliothek, Niedersächsische
Landesbibliothek.

The front of Westminster Abbey, where Newton was interred with great fanfare on March 28, 1726.

PHOTOGRAPHER: Jason S. Bardi.

The final resting place of Leibniz's remains is at this church in Hanover, Germany.

PHOTOGRAPHER: Jason S. Bardi.

Love it or hate it, Newton's work spread throughout the seventeenth-century intellectual world in a completely modern fashion. Reviews appeared in the literature summarizing, praising, and sometimes critiquing it, which was how Leibniz first came to find out about the book, in fact. He read a long review of it in the June 1688 issue of *Acta Eruditorum*, which praised Newton as "a distinguished mathematician of our time." The review was twelve pages of dry summary.

In a letter he wrote to a friend, Otto Mencke, in 1688, Leibniz said that he had been traveling and had not gotten his publications of late. But, he wrote, he had received a letter from a friend with a review of Newton's *Principia.* "I came across an account of the celebrated Isaac Newton's Mathematical Principles of Nature," he said. "That remarkable man is one of the few who have advanced the frontiers of the sciences."

High praise notwithstanding, Leibniz was reluctant to acknowledge the merits of the theory of universal gravitation as presented in the *Principia.* His own general view was that, while such phenomena as planetary motion could be explained mechanically and mathematically, the laws that governed them must arise from higher reasons. These higher reasons, he believed, were intelligible and logical, and, for him, the vortex theory that Newton was overthrowing made more sense.

Leibniz did do some interesting work in dynamics. He had postulated what is essentially the conservation of potential and kinetic energy—that, for instance, a ball held from a few feet above the ground and dropped will strike the ground with a kinetic energy equal to the potential energy it had by virtue of its position a few feet above the ground. Because he had done a great deal of thinking about some of the same sorts of problems that Newton had tackled, Leibniz was inspired to write three papers of his own on physical subjects, a few years after the *Principia* appeared.

One of these was his defense of the vortex theory, in his "Essay concerning the causes of the motions of the Heavenly Bodies," published in the *Acta Eruditorum*. In this work, Leibniz describes planetary motion in terms of harmonic vortices in which the sun is at the center of the world system, and the planets are carried around in the vortex. Because the planets were all revolving in the same plane around the sun, he couldn't think of a logical reason why they would do this if not for the existence of something like a vortex medium in which planets were spinning while they were carried around the sun.

Another of the papers he wrote after hearing about Newton's book was on a physical problem involving the resistance of a medium to motion. He used the publication concerning the problem of resistance of the medium to promote the ease with which such problems could be solved through his calculus. Leibniz had begun to feel enthusiastic about the possible applications of calculus, especially after a 1690 paper by Jacob Bernoulli appeared. Bernoulli's was an important document because it was written more accessibly than Leibniz's own papers, and it was the first in a long series that applied calculus to solving problems in mathematics.

One thing was sure, for Leibniz. He would be paying attention to the goings-on in England more closely for the rest of his life. In the fall of 1690, he sent a letter to a German ambassador in London, asking him to send news of discoveries and publications there. He had not gotten the *Philosophical Transactions* since 1678.

LIKE MOST OF the rest of Europe, Leibniz also was consumed with the dramatic events unfolding along France's eastern border. Europe was in turmoil in the late 1680s just when he was publishing his calculus papers and Newton was preparing to publish the *Principia*. At the end of the Franco-Dutch war in 1678, which Leibniz had crafted his Egyptian plan to avert, Holland was left free of French dominion but Louis XIV retained the Lorraine region. The king kept the latter militarily occupied, so he was poised in position to invade Holland or Germany again, should he choose to do so. And Louis was not the type to leave his troops idle forever—something that prompted Leibniz in 1683 to write a political satire, *Mars Christianissimus* (Most Christian War God), in which he called the French king the most powerful person in the world aside from the devil.

But for many in France, the actions of the most Christian war god were anything but a laughing matter. In the years leading up to 1685, a series of laws were passed in France that shut Protestants out of certain careers and encouraged the children of Protestants to declare their allegiance to Catholicism and be brought up as wards of the king. In addition, for the previous two decades, a number of French Protestants—Huguenots—had accepted an official government offer to convert to Catholicism and be exempt from taxes. Many more publicly converted to Catholicism because Louis XIV had exerted military force to influence their decision.

Soon thereafter, things went from bad to worse for Protestants in France. On October 18, 1685, Louis signed an edict that basically suspended all civil rights for the Huguenots, and the ripples that emanated from this chilling decree were profound. The edict ordered the demolition of Protestant chapels, called for an end to Protestant practices, closed Protestant schools, forced the baptism of Protestant children and authorized them to become the wards of the local judges, and allowed for the exile of pastors though not their flocks. This led to an exodus of as many as 200,000 refugees who fled France to seek asylum in Protestant countries. French Huguenots immigrated to England by the tens of thousands.

Meanwhile, a parallel political situation in England was throwing that country in turmoil. After several years of exile in France, Charles II had returned and become king of England on May 8, 1660. He rode into London, wearing his courtly finery, on May 29, 1660—his thirtieth birthday. The locals lined the streets and cheered. He had left England in defeat and now he returned in triumph, proving that if you cannot count on your own abilities, you may very well be able to count on the ineptitude of others—in this case, those who had followed Oliver Cromwell.

Voltaire described Charles II as having "a French mistress, French manners, and above all French money." Charles II was nicknamed the Merry Monarch because of his wit, charm, and love of good cheer. But he had a funny way of showing it at times. One of his first acts was to order executed ten people who had been involved in the trial of his father Charles I more than a decade before. Also, Oliver Cromwell was treated to the ironic insult of a posthumous execution. His cadaver was exhumed, and he was hung, drawn, and quartered, and dragged through the streets of London. Cromwell's head was placed on a pole in front of Westminster Abbey for the next fifteen years, until Charles II himself died.

Despite such demonstrations of pique, over the long run Charles II proved himself an amazing politician. He accepted kickbacks from Louis XIV because that king hated the Whigs (Puritans) so much that he was glad to pay Charles a salary while the latter ruled as king of England. And Charles very effectively kept the whigs in check throughout his reign. He even was able to dissolve parliament and have many of his Whig opponents arrested, without a civil war breaking out.

But when Charles II's son James—a Catholic—came to power in 1685, the stability vanished. For years, many Puritans in England had tried to get Charles II to disinherit his son—to bar him from ascension—on the basis of his faith, but the king never did. And when he came to power, James II was confident in living out the life he believed he was born to do—to rule his country. Upon becoming king, he met with his council and declared, "I have often heretofore ventured my life in defense of this nation; and I shall go as far as any man in preserving it in all its just rights and liberties."

In three years' time, James would flee England without fighting to keep his throne, and he may have become feebleminded in his later years because of syphilis. Whether or not he suffered from this communicable disease, one thing is certain regarding James: He was a disaster of a king. He had judges declare him the right to suspend laws at will, which he did—especially laws that curbed the power of

Catholics. He also raised a substantial number of Catholic soldiers from Ireland and stationed many of them near London, which was an infuriating if not frightening deed for the largely non-Catholic capital.

In the midst of all this, the *Principia* appeared, and in the front matter of the book was a dedication to James II as king. A year after that, James II was forced to flee England for good. This wasn't exactly James's fault. All his military and civil leaders abandoned him. Even his escape was ill fated. He was captured on his way to France—not by the English navy, nor the English army, nor the forces of the Dutch, but by salty, stinking fisherman. He escaped again, his second escape apparently helped by the fact that he was allowed to escape.

Nevertheless James had nobody to blame but himself, ultimately, since he alienated his allies as effectively as he did his enemies; he managed to unite the Tories and the Whigs against him. This inspired a number of whigs to sent a letter to Prince William of Orange in the Netherlands, in 1688, inviting him to become king.

By 1688, Louis XIV had been creeping toward war for several years. Ironically, his justification for his declaration of war was that the Ottoman Empire, he claimed, was planning to attack France and that Europe's eastern borders were not secure. So instead of attacking the Ottoman Empire to avert war in Europe, as Leibniz and Boineburg had proposed sixteen years before, Louis XIV attacked Europe to avert war with the Ottomans. The French king was very much opposed to William's taking over the throne of England because the two were adversaries in more ways than one.

William was an active leader in European resistance to Louis XIV, and, in 1686, he urged the reorganization of the Grand Alliance he had created in 1672 of the Dutch, the Holy Roman Empire, Spain, and Brandenburg. He created the League of Augsburg, whose members were the Holy Roman Empire, Spain, Holland, Sweden, Saxony, Bavaria, Savoy, and eventually England. The League of Augsburg was formed as an alliance against France after French troops invaded a German state and declared war against the Holy Roman Empire.

Louis threatened to declare war on England if William went there. But William, calculating that Louis was too busy invading Palatinate in Germany to make good on his threat, sailed to England with 15,000 men, landed in November 1688, and took the throne a few months later. As the third William to rule England, he became William III. The first was perhaps the most famous— William the conqueror, the first Norman king who had ruled the realm centuries before.

Unlike his renowned namesake, William III did not arrive the victor of a hard-fought conquest. He led an army that landed in England and deposed the king without a shot being fired (James II fleeing to France rather than facing war). Nevertheless, William looked the part of the gallant conqueror. A portrait of him by Sir Peter Lely, which resides today in the National Portrait Gallery in London, shows him in a suit of shiny black armor. A portrait of his wife, Mary, hangs nearby. She has wild brown hair and a gorgeous orange and crimson dress. They must have been a striking couple.

On February 13, 1689, England had its first double coronation— of William and Mary. Because William and Mary's coronation was unique in the sense that the two were crowned at the same time, they needed, for the first time, a second coronation chair. Apparently this chair had to be set lower than the other coronation chair because Mary was taller than her husband when seated.

It was a strange transition. William III was the deposed King James's nephew, and his wife Mary was James's daughter. William and Mary ruled England from 1689 to 1702, but the glorious revolution of 1688, as it was called, seriously curbed the power of the crown. William agreed to become king and his wife became queen, but they had to agree to a bill of rights that seriously curbed the power of the monarchy and established parliament as the rulers of the realm. Even though the parliament that emerged after 1688 was not representative of the people in the sense that it is today (then being controlled largely by the elite landowners, merchants, and nobles), it was nevertheless a stepping-stone to modern forms of government.

Moreover, England was now not only Protestant again but ruled by a king with a very personal interest in checking France's aggression toward Europe. These were strange times. French Huguenots fought with the English against their native countrymen, and English Jacobites (supporters of James II) joined the French to oppose the English.

Britain won a number of military victories against the French in the coming years. The English navy defeated the French fleet in 1692, and the war dragged on for another half-decade on land. The 1690s were a terrible time of war in Europe, and things were not helped by poor harvests, famine, and all the social problems that these spurred on. Against this backdrop, Leibniz and Newton were moving invariably toward a war of their own.

FOR NEWTON, THE *Principia* was a turning point in his life. It gave him the confidence to write the text that would become *Opticks* later. Meanwhile, demand fueled work toward a second and then third edition of the *Principia*, and he carried on extensive (perhaps even neurotic) correspondence helping others correct, revise, expound, and improve it—work that occupied Newton part time for most of the rest of his life.

As the editions of the *Principia* grew, so did the legend of Newton—and the popularizations of his science both profound and profane. A good example of the latter appeared in 1739, when a book, *Sir Isaac Newton's Philosophy Explain'd for the Use of the Ladies*, was published. An Italian named Francesco Algarotti was the author, and he lauded his own efforts for bringing a new kind of amusement to the ladies of the continent, whom he felt should be obliged to thank him. "If I have brought into Italy a new mode of cultivating the mind, rather than the present momentary fashion of adjusting their head dress and placing their curls."

Fame for his science aside, the *Principia* really changed Newton's life. Just after it was published, he was elected to parliament, a post that brought him to London. And this led him to meet Christian

Huygens, Leibniz's old mentor, who visited London for the first time in the late 1680s. This meeting was a significant one not only because it brought together these two stellar intellectuals, but also because it would introduce Newton to a young mathematician and astronomer, Nicolas Fatio de Duiller, a Swiss national who lived for several years in London and would play a crucial role in Newton's life.

Fatio is a fascinating character, and is a key player in the calculus wars. He entered the lives of Leibniz and Newton separately (the latter, in a most peculiar way), and was really the first person to stir up trouble between them.

Born in Basel, Switzerland, on February 16, 1664, Fatio was the bright son of a wealthy Swiss family. He went to Paris in the early 1680s to be educated, with a generous allowance and leave to study anything he wished. His father had made several attempts to study divinity, but Fatio chose to pick up mathematics and astronomy instead, for which he showed a great propensity, though his real talent in his early years seemed to be having the ability to be in the right place at the right time.

After Paris, Fatio went to the Hague to study, and at that point, as a young man of twenty-one, he met a certain Count Fenil. Fenil had been working as a military officer in France, when he shot dead his commanding officer and subsequently had to flee the country. He stayed at Fatio's home for a while, during which time he confided in Fatio a plot he was hatching.

As a way of making amends, Fenil had proposed to France's minister of war, the Marquis de Louvois, that he would seize William of Orange, then still the Dutch prince, and deliver him to King Louis XIV and France. The marquis took the bait. He sent Fenil a letter approving of the plot, promising a full pardon if Fenil succeeded and offering to pay for the operation. It was to be an ambush kidnapping raid. Prince William liked to take walks at the beach at Scheveling, about three miles from the Hague, where he lived. Count Fenil proposed to steer a light ship through the surf, land it in the shallows, hit the beach with about a dozen men, grab up the prince, and sail off to Dunkirk with him.

Bold as it was, the plan failed. Fenil's only mistake was to tell Fatio—and Fatio immediately told the plan to a doctor who was traveling to Holland, who passed word on to William.

This won Fatio the favor of the Dutch court, and he was rewarded for his disloyalty to Count Fenil with the promise of a professorship of mathematics at the university at The Hague, with a nice salary and the comfy job of instructing nobles and the gentry. However, while these arrangements were being made, Fatio went to England, where he eventually became a mathematics teacher in Spitalfields. He made a few trips home in the 1690s, but otherwise resided in England for most of the rest of his life.

When Fatio arrived London in 1687, he wasted no time establishing himself among the British scientific elite, managing to get himself elected to the Royal Society in just two weeks. Newton had just published the *Principia*, and it was the talk of town in the circles Fatio frequented.

Huygens came to London that summer, and Fatio, exploiting his relationship with William of Orange, won the right to escort the famous older scientist around town. Huygens and he hit it off, and they became friends. Huygens became something of a mathematical mentor for Fatio, much as he had been for Leibniz. Then Fatio was introduced to and charmed one of the most important men he would ever meet—Isaac Newton. Escorting Huygens brought Fatio into contact with the Englishman.

Fatio had become infatuated with Newton's theory of gravitation from the moment he arrived in London, and he became friends with Newton after the two met at a meeting of the Royal Society that they both attended on June 12, 1689. Newton had come to the meeting to meet Huygens, and Fatio was there with Huygens, but the real connection was between Fatio and Newton.

Their intense friendship in the early 1690s is the source of some historical speculation. Newton's letters to Fatio are unusually warm, and some have suggested that Fatio was the object of Newton's latent homosexual affections. It is more than tempting—indeed fun—to read the Englishman's close relationship with his young protégé as one that reveals the root cause of his affection and to try to read between the lines of their correspondence. For his part, Fatio wrote to Huygens that he was "frozen stiff" when he saw what Newton had accomplished. Likewise, one might raise an eyebrow at the reports that Newton liked to build doll furniture and preferred the company of girls—suggesting that his preference for girls (as opposed to adult women) might have belied a preference for boys.

However, there is scant historical evidence that Newton had any interest in either sex. According to Voltaire, Newton died a virgin after more than eighty years on Earth—a virtue, Voltaire adds. For Newton, sex may have been as enticing as a tray of pudding and tea left outside the door of his study by his servant—the pudding goes cold and uneaten on one side of the door while Newton works all night scribbling strange symbols in notebooks on the other.

Whatever their relationship, the two were unusually close and were great admirers of each other, as is evidenced from their correspondence, which makes it clear that Fatio was very fond of Newton. Within a few months of their first meeting, Fatio wrote a letter to his friend Jean-Robert Choet, calling Newton the most honest man he knew and the ablest mathematician who ever lived. Fatio offered to sit with Newton and help him read a new book that Huygens had just published (in French).

The passion of their friendship was mutual. The earliest letter between the two was written by Newton later that year, on October 10. Newton wrote to Fatio and asked if there would be any rooms at the Swiss's residence in London. "I intend to be in London the next week," Newton wrote. "And I should be very glad to be in the same lodgings with you."

Over the next two years, Fatio and Newton became closer and closer. Even when Fatio left England for fifteen months in June 1690, Newton had him on his mind, writing to John Locke, for instance, on October 28, 1690, "I suppose Mr. Fatio is in Holland for I have heard nothing from him the half year." When Fatio returned in

September 1691, Newton rushed to London to meet him in private as soon as he was back and, after that, their relationship became all the more intimate. They were frequently seen together in the Royal Society in London, so much so that, when their presence was recorded in the attendance notes, they were often marked down as a single unit. Hooke, still Newton's nemesis in the 1690s, began calling Fatio "Newton's ape."

Fatio fancied himself as more than Newton's ape, and he offered to supervise the revisions of the *Principia* to make the second edition of the book. He envisioned his role as something approaching Newton's collaborator, and he wrote to Huygens that this second edition would be much longer because of his additions.

If Newton got along famously with Fatio, Leibniz had a strange relationship with him—nothing like the mutual admiration society Fatio formed with Newton. Huygens tried to get Leibniz and Fatio to correspond, but the German didn't see the need. Leibniz was already serving as a mentor for a growing cadre of European mathematical intellectuals—and Fatio would not be one of them.

Calculus was also already on the move. In 1691, Johann Bernoulli went to Paris and became the teacher of the Marquis de L'Hôpital. This was a fruitful connection because L'Hôpital would write a few years later, in 1696, one of the first ever calculus textbooks, *Analyse des Infiniment Petits* (Analysis by Infinitely Small Quantities), with a great deal of help from Bernoulli.

Leibniz was on the move as well.

$\boxed{8}$

The Shortest Possible Descent

■ **1690–1696** ■

Men act like beasts insofar as the succession of their perceptions is due
to the principle of memory alone . . .

—Leibniz, *Monadology*, published in 1720

It's nighttime somewhere along the coast of Italy in the last decade of the seventeenth century, and a small ship bobs upon the Adriatic Sea with a small crew and few passengers, including one foreigner with courtly German manners and a quiet demeanor. The crew is worried. A storm, a storm is blowing! The ship is tossed, and all aboard are shocked, discomforted, fearful. The ship's crew is probably cursing in five different languages before one of them finally says in Italian to his fellow sailors that the cause of the storm is their German passenger—a Protestant!

That Lutheran Judas has brought the wrath of God upon us! Throw him overboard! Throw him overboard!

But they note that the stranger is sitting quietly, like the calm eye at the center of a storm, passing something through his hand. What is it? A rosary?! Look at him praying! He must be a true believer. Let him live . . . He is one of us.

Strange as it sounds, the basis of this tale is true. To Catholics three hundred years ago, a rosary in the hands of a Protestant traveler was like a Canadian Flag on an American traveler today. Leibniz saved himself from murder at the hands of the superstitious sailors ferrying him between towns in Italy because, unbeknownst to the crew, he could understood enough Italian to know he was in danger unless he feigned Catholicism double-quick.

The event is one of the most interesting from a long trip that he took through Germany and Italy from the autumn of 1687 until the summer of 1690, to do the research he needed to write a short history of Ernst August's family, the House of Brunswick-Lüneburg, which Leibniz had proposed doing a few years earlier within a few weeks of the failure of the project to drain the mines at the Harz Mountains.

Such histories were common in those days because the fortunes of the state ultimately depended on the fortunes of the noble heads of state. Nobility was hereditary, pedigree of utmost importance, and so genealogies became an important way of justifying, if not furthering, the sociopolitical positions of Europe's leaders. In the seventeenth century, many scholars hired themselves out to noble patrons to research such family histories, often tracing the family back through the centuries to the Middle Ages or even earlier.

Because so much was at stake, flattery would often supplant history, as noblemen and women were often mapped back in lineage to Charlemagne, who by the seventeenth century must have had more descendants on paper than Genghis Khan. Many of these works were downright ridiculous. A Venetian theologian even claimed he could trace the royal Habsburg family back to Noah's ark. Ernst August himself once received a bit of this kind of flattery from a Dutch nobleman who traced his line through Augustus Caesar all the way back to Romulus and Remus.

The duke was not so foolish as to believe this all this, but it sparked in him an interest to learn the real history of his family. Other historians had asserted that the family was related to the House of Este, one of the oldest and most noble in Europe. If this were true, then it would lend a great deal of credibility to Ernst August's ambition of furthering the fortunes of his family. In those days, one of the best ways of doing this was to show a noble pedigree, which in his case would have meant "Estefication," but nobody had ever been able to prove the noble Brunswicks were related to the Estes.

Leibniz had a very realistic goal: He wanted to trace the family back about one thousand years, to AD 600, and fill in all the gaps in between. But to do this he would need to travel widely in search of sources in state archives and monasteries strewn across Germany and Italy; there was no way he could accomplish this by staying within Hanover. As soon as the mine project ailed and he was ordered to cease, Leibniz began petitioning Ernst August to allow him to undertake the research. He was not just seeking permission to travel and write, but paid expenses and a secretary.

Ernst August was sufficiently impressed by the proposal to approve Leibniz's plans, appoint him court historian, and authorize him to research and write the history. This was a major coup for Leibniz. Finally he could travel, study, write, meet, and correspond with other scholars without having to worry about where his money was coming from.

Leibniz departed in the fall of 1687, and for the next two and a half years went to cities all over Germany, Italy, and throughout southern Europe: Bologna, Dresden, Frankfort, Florence, Marburg, Modena, Munich, Naples, Padua, Parma, Prague, Rome, and Vienna. Indeed, Leibniz was to happily indulge in this kind of travel for most of the rest of his life. He did so much traveling, away from his home base for weeks, months, or years at a time, that he designed and commissioned a folding leather chair to accompany him so that he would have a place to work wherever he went. This ornately designed chair had a seam that ran down the middle, the bottom struts hinged so that it could fold easily. This invention was characteristic of Leibniz—he was constantly trying to adapt the world to fit his desires or needs, and was not solely interested in how things were but how they could be. He lived a life that was not fit for the world at times, so he made his world fit for him.

On his travels, he took many detours and took in many sights, for instance climbing to the top of Mount Vesuvius in Naples and exploring the catacombs in Rome. He also met many people and discussed many subjects that were unrelated to his purpose—something that he was wonderfully happy about, as he indicated a letter to Antoine Arnauld: "As this journey has served to free me in part from my ordinary occupations, and to furnish my mind with recreation, so have I had the satisfaction of engaging in conversation with many gifted persons respecting science and learning."

When he was in Rome, Pope Innocent XI died, and Leibniz schmoozed with the cardinals who came from France for a conclave. When a new pope, Alexander XIII, was chosen, Leibniz conscientiously wrote a long poem hailing him.

Several of the people he met, he would later correspond with for years to come. For instance, he met a Jesuit priest, Claudius Philip Grimaldi, who was about to depart to go to China as a missionary. Leibniz was very interested in Chinese things, and he believed that the Chinese language was based on a profound philosophy that had been forgotten—even by the Chinese themselves. For the rest of his life Leibniz was to have a singular passion on matters related to China and a cultural exchange between east and west, and so he relished his correspondence with Grimaldi.

Leibniz met the celebrated Italian doctor Bernardino Ramazzini, who has been called the father of industrial medicine. They held each other in very high esteem. Leibniz was a strong advocate of health care, and he believed it was the moral duty of governments to provide it. He heavily promoted a cure for dysentery that was found in a root from South America. Leibniz also advocated preventative medicine, once writing a memorandum in 1681 prompting military health through such peacetime activities as sports. He proposed the idea of health councils and strongly advocated the isolation of cases during disease outbreaks to curtail epidemics. After being encouraged by Leibniz, Ramazzini produced a statistical record about health in 1690 that was championed by Leibniz in Vienna and to some of his acquaintances in France.

Upon a trip to Vienna, Leibniz was given his first audience with the Holy Roman emperor, and he took the opportunity to pitch a wild ride of ideas, following them up with several memoranda. He was exploding with ideas: a tax on fancy clothes; street lamps for the city of Vienna (which as it happens was eventually carried out); central archives and libraries; major economic reforms; and ways to improve manufacturing.

The history of the House of Brunswick was in a certain way a tremendous success, and Leibniz was able to deliver on his promise to research the origins of the duke's family. There had been a hypothesis that a marriage between a family in northern Italy and one in Bavaria had taken place several centuries earlier, which had involved one of the duke's ancestors in the Guelf House. Following up on this, Leibniz tracked down old Este monuments and, in 1689, found a tomb in Modena that was engraved with the names of the deceased. He also located papers supporting the families' legal connection, and so the physical and ephemeral evidence together served as a fairly good verification that the marriage had indeed taken place. By the beginning of 1690, he had pored over enough documents to make Milton go blind a second time, and was proud to report to Ernst August that he had firmly established the the relationship of the duke's family with the House of Este.

This effectively increased the prestige of the House of Brunswick, ultimately enabling the elevation of the Hanoveran dukes to the electorate of the Holy Roman Empire—one of the handful of German nobles who could vote for the Holy Roman emperor. (The emperor had been chosen, since the year 1356, by certain German princes who were known as "electors" and who fancied themselves heirs to

the glory of the Roman senate among the plebian mishmash of the
350 other, mostly smaller, political entities that composed the Holy
Roman Empire.)

Elevating Ernst August to elector was not a straightforward matter, as several of the other German princes opposed the move for a
variety of reasons. Leibniz wrote a number of papers to support the
Brunswicks' cause, based on historical analysis, legal precedent, and
diplomatic arm-wrangling. In all, for eight years, beginning in 1684,
he was to work hard behind the scenes on the negotiations for the
new electorate. Finally, in 1692, Ernst August achieved his ambitions
and was made an elector, an honor his heirs would thereafter inherit
without any need other than birthright for qualification.

In 1696, Leibniz was promoted to privy counselor of justice—an
office of high rank probably awarded, at least in part, for his involve-
ment in raising the duke to an elector. This resulted in the addition
of a bonus to his salary.

From this perspective, Leibniz's trip was a smashing success. Had
he been able to stop working on the history at the point he established the connection between the duke and the House of Este, the
project would have been a total success. But this was something he
could not do. He still had to actually *write* the history of the House
of Brunswick. Substantiating the Este connection had been only part
of the deal. In January 1691, a year after he wrote to Ernst August
from Italy, telling him the good news that he had established the
favorable family tree, Leibniz now prepared an outline and presented it to the duke, saying that he estimated the history might take
two years to write.

He had no idea what he was in for.

The project was a major undertaking and, even with the help of
assistants, Leibniz was never able to finish it. In fact, the history
dogged him for the rest of his life, and there was little in his later years
that was not clouded by its incomplete assignment. The assignment
took time away from his other studies in mathematics, physics, and
philosophy, and when he was on his deathbed, in the throes of the

calculus wars with Newton, the history was still hanging over his head like a paper sickle.

He expressed his frustration well in 1695, in a letter to a man named Vincent Placcius: "I cannot describe to you how distracted a life I am leading. I search for different things in the archives and look over old papers and manuscripts never printed, hoping to get some light respecting the history of the House of Brunswick. Letters I receive and answer in great numbers. But I have so much that is new in the mathematics, so many thoughts in philosophy, so numerous literary observations of other kinds, which I do not wish to lose, that I am often at a loss what to do first, and feel the truth of Ovid's exclamation, *Inopem me copia fecit* [plenty has made me poor]."

In 1696, a premature report of Leibniz's death was circulating in England. Hearing of this, Leibniz wrote a letter to Thomas Burnet in England, complaining about how busy he was: "If death will only grant me the time requisite for the execution of the works already projected by me, I will promise to enter upon no new undertaking, and industriously to prosecute the old ones; and even such an agreement would defer the end of life no inconsiderable period."

Unfortunately, there would be no respite—more than twenty-five years later, Leibniz died still working on the bloody thing. It became his *opus tedium* and, later in life, he wrote to mathematician Adam Kochanski that the history was his Sisyphean stone to which he was bound. When Leibniz died, he had only gotten as far as the year 1005, and it would not be until more than a century after he died, that the history was finally published in three volumes.

Perhaps the reason for Bertrand Russell's lament that much of Leibniz's time in service of the dukes was a waste of time stemmed from the untold hours the German mathematician had spent during some of his most productive years, working on what seems now to be not only a mammoth but also pointless exercise in genealogical research. It's true that establishing the Este connection helped in the elevation of the duke to electoral status, but the rest of the genealogical research did not do much to contribute to the family's ultimate improvement—Ernst August's son, George Ludwig, being elevated to a regent, which would happen when Leibniz's third boss, this son, became George I, king of England, in 1714.

The decision to make George king came not from the ancient pedigree of the family but because of more recent ancestry and a solid Protestant pedigree. He was the great-grandson of England's King James I, but, more important, he was thoroughly Protestant. And when he assumed the throne, what should have been a happy time for Leibniz—as one who was nominally in the inner circle of the court at Hanover—was in fact bitter. The writing of the Brunswick family history kept Leibniz away from the new court because George Ludwig used it as an excuse to not allow Leibniz to accompany him to England.

Even so, at least one interesting thing did emerge from the project. The preface, the *Protogaea*, which he wrote in 1693, was a fascinating natural history of the Earth and the region where the duke and his ancestors lived. In it, Leibniz delved into prehistory, going back before human creation.

In his *Protogaea*, Leibniz proposed that the planet was originally hot, and that it had cooled, formed a crust, and then water had condensed on its surface. He explained the influence of volcanic activity on geological history and sedimentation, discussed fossils, and anticipated Darwin's theory of evolution by proposing that the earliest animals were marine and that land animals came later. One nineteenth-century commentator notes that the *Protogaea* contains "the germ of some of the most enlightened speculations of modern geology."

Leibniz published an account of his *Protogaea* in the *Acta Eruditorum* in 1693, but the essay itself was not published until after his death. Some writers have suggested that the project was the perfect example of a Leibnizian endeavor. "The mode in which he prosecuted his task, the immense gyrations of thought in which he indulged, the number of subjects which were successively taken up, the eagerness with which he pursued each, the gigantic scale on which he framed his plan, and not least of all, the scanty fragments he left of the whole, are so remarkably characteristic of his genius and his habits."

―――――

NEWTON WAS ALSO experiencing the utmost highs and lows that his life had to offer, during the last decade of the seventeenth century. The *Principia* had been well received, and the year after it was printed, he had been elected to Parliament as a representative of Cambridge University. This election, which brought him to London, gave him a taste for public service that he would never lose. Newton also began to lobby his friends and contacts for a permanent administrative job. He tried to get John Locke to get him a position at the Mint in 1691, and another friend tried to get him a position at King's College, London. His friend Charles Montague, who came to Trinity in 1679 and knew Newton's genius firsthand, was also enlisted to get a post for Newton. Though he was unsuccessful at first, Montague ultimately did help to secure a governmental post for him.

In 1693, some of Newton's mathematical work finally made it into print for the first time. He did not publish this work himself, but rather allowed John Wallis to publish it within some volumes of Wallis's own mathematics. Wallis was a charming man and a brilliant mathematician, though perhaps a little flawed as well since he was first and foremost a *British* mathematician and went out of his way to promote the supremacy of British accomplishments. In Wallis's 1693 and 1695 books, he devoted pages to Newton's contributions, and compared fluents and fluxions to the calculus that Leibniz had published a few years earlier. "Here is set out Newton's method of fluxions, to give it his name, which is of a similar nature with the differential calculus of Leibniz, to use his name for it," Wallis wrote, "as anyone comparing the methods will observe well enough, though they employ different notations. . . . "

Wallis also referred to Newton letters of 1676 and said that in them Newton explained his methods to the German mathematician. This is a significant moment in the calculus wars because, in going over these passages, many of Wallis's readers were for the first time encountering the notion that Newton had developed methods that, lo and behold, had actually preceded and were identical to Leibniz's calculus. And with this revelation came the first suggestion that one man's work was *better* than the other's, because the much-respected Wallis championed the ease of Newton's fluxions and fluents over Leibniz's calculus. In one passage, for instance, Wallis wrote, "And although at first glance fluents and their fluxions seem difficult to grasp, since it is usually a hard matter to understand new ideas; yet he thinks the notion of them quickly becomes more familiar than does the notion of moments or least parts or infinitely little differences." This claim did not do much to turn opinions away from Leibniz, but it was really the first salvo.

Some of the book's readers on the continent were astounded by its claims regarding calculus. After all, Leibniz's papers on calculus had been read all over Europe, and since Leibniz never mentioned in them anything about Newton, many Europeans didn't know what to make of the subsequent British publication of Newton's earlier work. They still had not seen anything of Newton's methods that Wallis was touting. Nor could any of these methods be found in print anywhere. Leibniz's calculus, on the other hand, had been in print for a decade, and it was really starting to bear fruit, with the Bernoulli brothers and others learning, developing, and beginning to apply the methods to solving complicated problems.

Johann Bernoulli was a little peeved at what he saw as a slight against Leibniz, after he read the relevant sections in Wallis's books. He wrote to Leibniz saying as much. Leibniz took the higher ground at this point. "It must be admitted that the man is outstanding," he wrote to Bernoulli. But Bernoulli regarded Newton's work as derivative of Leibniz's, and he was so blunt as to suggest the possibility of plagiarism—that Newton had borrowed his ideas from Leibniz: "I do not know whether or not Newton contrived his own method after having seen your calculus, especially as I see that you imparted your calculus to him, before he had published his method."

Leibniz was not silent about Wallis's books and their treatment of his mathematics. He wrote a letter to Thomas Burnet complaining, "I am very satisfied with Mr. Newton, but not for Mr. Wallis who treats me a little coldly in his last [volume of] works in Latin, through an amusing affectation of attributing everything to his own nation." In this letter, he employed a device that he would use for the next two decades every time the issue of calculus came up. Ask Newton, Leibniz essentially said. Newton knows—he'll tell you. And Leibniz apparently took Newton's apparent silence on the matter as an acknowledgment that Leibniz was within his rights to claim his own independent invention of calculus.

Interestingly, Wallis was a much better ally to Leibniz than to Newton. Wallis was not out to get Leibniz, but saw him as a legitimately esteemed mathematician who made an independent discovery of calculus, and, in the last few years of Wallis's life, he and Leibniz exchanged some eighty pages of letters. In actuality, Newton may himself have ghostwritten the personally praiseful passages that appeared in Wallis's book; at the very least, Wallis was writing what Newton wanted him to write and, two decades later, when the calculus wars came to a head, it would be Newton who would point to Wallis for support of his case: Ask Wallis. He knew.

At this point Leibniz, for his part, was perfectly willing to put on a good public face and give Newton his due credit. In the early 1690s, Leibniz and Huygens were in communication again, until the latter's death in 1695 cut short their renewed correspondence. Huygens wrote a letter to Leibniz after he had seen a volume of Wallis's *Algebra*. In it, he told Leibniz, he had encountered "differential equations very much like yours, apart from the symbols." Leibniz obtained extracts of Wallis's book from Huygens in 1694, and after reading what was written of Newton's work, wrote to Huygens saying, "I see that his calculus agrees with mine," but adding that his own methods were "more fitted for enlightening the mind."

Even so, there is some evidence that Leibniz didn't want things to stand at that. An anonymous review of Wallis's work appeared in the scholarly journal *Acta eruditorum* that treated Newton's work as if it were merely a celebration of Leibniz's skills as a mathematician. This review was most likely written by Leibniz himself, who loved to write anonymous scientific letters in which he both attacked other mathematicians and praised himself. (He once wrote an anonymous review of some of his own work in which he referred to himself as "the illustrious Leibniz.")

Meanwhile Leibniz probably didn't really see Newton as a threat because he was seeing tremendous success in his own intellectual endeavors—he was at the top of his game. Finally, in 1694, Leibniz had found a skilled artisan to help him perfect a working model of his calculating machine, which could multiply numbers up to twelve digits.

Leibniz published the fullest account of his philosophy in 1695 in a French journal under the title "Systéme nouveau de la nature et de la communication des substances." It was his account of his metaphysics, which went all the way back to his logical studies as a college student and on which he had been working more or less continuously in the three decades since. This put Leibniz on the intellectual map of Europe. Many had already known of him through his vast correspondences and his various mathematical and philosophical papers, but the article really made his name well known. He became even more of a public persona for his philosophy after Frenchman Pierre Bayle wrote a dictionary and included a critique of Leibniz's work in it.

Mathematically, on the continent, Leibniz was the grandfather of calculus—its utmost authority. When L'Hôpital planned to write a calculus book in 1694, he first wrote to Leibniz about it, spelling out some of the problems he intended to solve.

Had Leibniz chosen to attack Newton during the last decade of the seventeenth cetury, he surely would have won the calculus wars. Newton was not yet in his position of maximum strength as the president of the Royal Society, and he may well have never recovered if Leibniz, then at the peak of his fame, had come after him. But Leibniz would not have done this, because he felt no malice toward

Newton at this point. He even wrote to Newton, in 1693, a letter full of praise and veneration for his esteemed colleague.

This brief exchange of letters that Newton and Leibniz made directly to each other was both friendly and meaningless. "How great I think the debt owed to you, by our knowledge of mathematics and of all nature, I have acknowledged in public also when occasion offered," Leibniz wrote, opening up the exchange. "You had given an astonishing development to geometry by your series; but when you published your work, the *Principia*, you showed that even what is not subject to the received analysis is an open book to you."

Leibniz added that he wished to see Newton continue with his studies of the mathematical nature of the world. "In this field you have by yourself with very few companions gained an immense return for your labor," he wrote.

In Newton's reply to Leibniz's letter some six months later, he paid his correspondent an incredible compliment as a person he regarded as one of the top mathematicians of the day: "I value your friendship very highly and have for many years back considered you as one of the leading geometers of this century, as I have also acknowledged on every occasion that offered." Also in that letter, Newton translated the anagram from the letter he had sent Leibniz in Paris two decades earlier, which the German was happy to finally receive. Clearly, Leibniz saw no need to challenge a seemingly chummy Newton at this time. He did not view him as a threat.

Johann Bernoulli, as one of Leibniz's loyal followers, was not so willing to let things go. Bernoilli hatched the idea that he would reveal Newton's inability to compete with his friend when it came to mathematics. In 1696, he issued a challenge called the "brachistochrone problem," and addressed it, with no small amount of gregariousness, "to the shrewdest mathematicians in the world." Individual copies were posted to Wallis and Newton in England, and Leibniz published an article on the problem in the German journal *Acta Eruditorum*, as well as had it advertised in the French *Journal des Sçauans*. Solutions were to be accepted up until the following Easter.

This sort of competition was one that Leibniz had established a few years earlier, when he issued such a challenge to Abbé Catelan in 1687. The problem was to find the curve along which a body would descend without friction and at a constant speed. Huygens, Leibniz, and the Bernoullis had all participated in it.

These sorts of problems served to demonstrate the power of calculus. Jacob Bernoulli had proposed a similar problem in 1690, and when Leibniz worked out the solution, he sent it to the *Journal des Sçavans* for publication in 1692. In his article, he touted the power of infinitesimal calculus to solve this and other problems with ease and speed. He sent another letter to the journal a few months later, and another in 1694 where he reiterated the power of calculus over Descartes' inferior analysis. He also praised Johann and Jacob Bernoulli for applying his calculus, mentioned L'Hôpital and his work, and, interestingly, wrote that Newton had a similar method but used inferior notation.

In 1696, Johann Bernoulli wanted to test just how powerful Newton's "similar" method was when he came up with the brachistochrone or "shortest time" problem. Bernoulli's challenge was to determine the curved line that connects two given points, one not directly beneath the other, along which a heavy body falling under the influence of gravity would descend in the shortest possible time. This is a classic example of the type of problem that calculus could solve—a problem for which a general solution can be found that expresses the curve without defining any specific parameters of the problem, such as the mass of the object or the distanced between the two points. And it was the ultimate challenge to test Newton's abilities, since only true masters of calculus could possibly solve it.

The problem was a painful one, as I recall from my encounter with it in a junior-level physics class that I took more than a decade ago. I remember spending most of a Saturday working toward a solution, but I couldn't get it right. A few days later, I showed up early to class and confessed to my professor that I hadn't been able to solve it after exhaustive efforts. "Don't feel so bad," said my professor; "three hundred years ago, there were only three or four mathematicians in the entire world who could solve it."

Actually only five mathematicians were able to solve the problem—or at least sent the solution back in the agreed upon time frame. These were, not surprisingly, perhaps the only five people on the planet who had mastered calculus: Leibniz, Newton, L'Hôpital, and the Bernoulli brothers. Newton, of course, had no problem solving it, and he did so with apparent ease. He received it on January 29, 1697, when, after working a full day at the mint, he came home and solved the problem in a single night, and sent his answer back to Bernoulli anonymously.

This fact was not missed on Leibniz, who gloated, "they only solved the problem whom I had guessed would be capable of solving it, as being those alone who had penetrated sufficiently deeply into the mystery of our differential calculus."

The gambit failed to ferret out Newton as one with less skill in mathematical analysis, but it proved the supremacy of calculus. For Leibniz, calculus was an elite club of which he was the founder. He was not threatened by the fact that there was another member—Newton—across the English channel who had apparently come up with his own independent methods and was able to apply them with great success. The Leibniz school of calculus was dominant and rising, and, to him, the Newton school was . . . a footnote, really.

If anything, Leibniz rather pitied the man. After all, his own rise to intellectual supremacy in the early 1690s Europe had coincided with Newton's deteriorating mental state. The British mathematician was not well, and rumors had spread through Europe that he had the worst possible illness a genius could have.

In 1693, Newton is reported to have had an almost complete mental breakdown, the cause of which has been the source of a great deal of historical speculation through the years. His symptoms, in modern terms, were insomnia, loss of appetite, memory loss, melancholia, and paranoid delusions. The delusions were manifested in letters he sent to his associates, and his insomnia and other sytomps are gleaned from those same sources. In one of these letters, he wrote that he had slept only nine hours over the course of two weeks. And he had refused food during that time as well.

Various reasons have been suggested for his illness, the most obvious symptom of which was his almost complete lack of sleep. Of the sleep, it must be said that Newton had been spending much of his time in excessive study, as he was wont to do, but even for a workaholic his sleeplessness was extreme.

Some have suggested that the lack of sleep was really just a symptom of a much deeper cause—Newton may have been suffering from chronic mercury poisoning. He certainly showed the symptoms—sleeplessness, digestive problems, loss of memory, and paranoid delusions. There is also no question that he was exposing himself to perhaps dangerous amounts of chemicals in the course of his alchemical experiments. In the late 1680s and early '90s, Newton made experiments on different alloys of iron, tin, antimony, bismuth, and lead. His notes indicate that he was finding ways to combine different amounts of metals into alloys. He found one alloy, for instance—two parts lead, three parts tin, and four parts bismuth—that melted in the summer sun.

However, the case for mercury poisoning was weakened by the lack of additional symptoms that one would expect to accompany mercury poisoning severe enough to cause insomnia, including hard-to-miss symptoms such as gastrointestinal problems, gingivitis, neurological deficits, and chronic fatigue.

At least one psychiatric professional has argued against the mercury poisoning and instead in favor of the idea that Newton's mental state was not toxic in nature, but rather psychological—that he suffered from manic depression (or bipolar disorder, as it is now called). Strong support of this theory may be in the fact that Newton seems to have suffered from insomnia many times in his life, which is consistent with manic depression and its tendency to manifest episodically throughout a person's life.

Other signs from Newton's childhood include things like the facts that he was often unkempt, a loner, shy, and did not seem to engage in any sort of recreational activity. His college years were marked by isolation, and manic depression may have been the root cause of many of the problems he had in his life—such as his battles with Hooke and Leibniz.

This analysis, while perfectly plausible, is impossible to prove. And it is far from the only theory out there. Another theory is that the mental breakdown was caused by a severe professional trauma that Newton suffered in 1692. According to legend, tragedy was a candle left burning and a window left open. One day in February of that year, he went to church and left a candle burning on his table. It somehow blew over without extinguishing and set fire to a ream of papers, including the sole copies of some of his valuable notes on optical experiments, physical observations, and other subjects that he had been perfecting for decades. Newton arrived home to discover the fruit of many years' labor had burned to crumbling black flakes of ash.

It's not clear how much of an impact the loss of these papers had, but the theory is that it may have been the cause of Newton's crumbling to the edge of sanity.. The loss of such an irreplaceable collection of notes comprising about half of his life's work would certainly have been a devastating blow to any scientist before our age of backup disks.

Another version of the same story has Newton's dog Diamond knocking the candle over onto the papers, again reducing them into ashes. In this account, Newton appears at the door like a swooning Southern belle with a British accent and laments, "Oh Diamond! Diamond! Thou little knowst the mischief done!"

As amusing as this latter scenario is to imagine, there is no evidence that Newton ever even had a dog named Diamond. The story may be no more accurate than the one about the fallen apple giving Newton the idea for universal gravitation. And there is evidence that the rumor of the fire itself was just that—a rumor. There was apparently a fire years earlier, in 1678, which had indeed burned some of his papers after Newton left a candle burning in the empty house, and there may have been some confusion about this when the rumor spread in the 1690s of a fire destroying a substantial quantity of his writings. In fact, the rumor became so overblown that at least one person reported that Newton's entire house had burned down.

Another theory is that, fire or no fire, his incapacity was more to do with his relationship with Fatio, which had been growing more and more intense in the months leading up to the breakdown. A dramatic turn in their relationship occurred when Fatio fell ill with pneumonia in 1692, after returning to London from Newton's Cambridge residence on November 17 of that year. "My head is something out of order, and I suspect will grow worse and worse," he wrote to Newton. Fatio went on to detail his symptoms—a congestion that felt bigger than his fist in his chest—and he said that he tried all the normal medicines and treatments to no avail. "I have Sir almost no hopes of seeing you again," Fatio wrote. "Were I in a lesser fever, I should tell you sir many things."

Newton wrote that he could not even express how much he was affected to hear of it. He offered Fatio money, and wanted to keep him in Cambridge and nurse him back to health. "For I believe this air will agree with you better," Newton responded, signing his reply "Your most affectionate and faithful friend to serve you, Isaac Newton," and he sent it special delivery to London to Fatio, who by that time had already almost recovered.

Nonetheless, a few months later, Fatio wrote that he would like to take Newton up on his offer and stay with him in Cambridge—especially if he could be able to stay in the rooms next to Newton: "I should be glad to know sir what prospect you had before you of a way for me to subsist at Cambridge."

Unfortunately, "The chamber next me is disposed of," answered Newton. Still, he again offered to give Fatio money, an allowance, whatever it would take to get him to stay near him in Cambridge and make his stay there easier. To this Fatio replied, "I could wish sir to live all my life, or the greatest part of it, with you, if it was possible."

Yet, instead, Fatio decided to leave England and return home to

Switzerland. After two final meetings in May and June of 1693, he dropped out of Newton's life—nearly for good. We will never know what passed between these two men. What we do know is that, in 1693, all intimate contact between the two came to an abrupt and final end, and this was about the time that Newton fell into a severe depression.

Whatever the cause of his madness, it manifested in strange ways. He sent disturbing letters to Samuel Pepys and John Locke saying that he had not slept or eaten in months; he wanted to cut off all correspondence with Montague, convinced that he was false; he apologized at length for minor snubs to which he had subjected Locke; and so on.

There is one final possibility to consider regarding Newton's condition: that he was not poisoned by a toxin, wracked with depression, or overwrought at the loss of a friendship at all, but was quite simply being Newton. His sleeplessness might not have been a symptom of some underlying neurological defect but rather an ordinary bout of restless energy, the likes of which fueled him at many times in his life. Likewise, paranoid anger, which is often listed as one of his primary symptoms, was something that characterized many of his relationships. Not exclusive to the 1690s, the famous Newton temper was to rear its ugly head throughout his life. He struck up a nasty fight with the astronomer John Flamsteed, for instance, convincing himself that Flamsteed was to blame for his not being able to come up with an adequate theory of the moon's motions. Newton had not been satisfied with lunar theory as it was laid out in the *Principia* as he wrote it in the 1680s, and he worked on improving it in subsequent years. In 1694, he began to use Flamsteed's observations to elucidate the moon's orbit. He worked on this on and off for several years.

This was to have been one of the first examples of what would become a standard sort of scientific collaboration between the theorist with the experimenter, the perfect marriage of theory and experiment. Newton, though himself a skilled experimenter, would act the theorist and apply his penetrating geometrical skills to the data sets that Flamsteed, the astronomer, would provide him with. The experiment, like the collaboration, failed—in large part because Newton was so overbearing that he spoiled their relationship.

But even if Newton did not have some sort of nervous break-down, the effect on Leibniz was the same. He heard that Newton had had . . . something, and he was sympathetic for the man, whose greatness he recognized. Leibniz had genuine concern for his British rival.

This concern cropped up again a few years later in 1695, when Burnet, then royal physician of Scotland, visited Hanover and befriended Leibniz. When Burnet returned to Britain, Leibniz kept in touch with him and used him to feel the pulse of life and events in London.

Actually, he relied on Burnet to keep tabs on Newton as well. After their brief exchange in 1693, Leibniz took an occasional interest in Newton and his affairs, and had at least one more occasion to get a note to him through Burnet in 1696. Burnet reported back that its recipient was gracious and thankful for the letter but busy because he had just become warden of the mint, a position Newton had been trying to get for some time.

Newton's friend Montague had written to him on March 19, 1696, with the good news of his appointment: "I am very glad that at last I can give you a good proof of my friendship, and the esteem the king has of your merits . . . the King has promised me to make [you] Warden of the Mint, the office is the most proper for you 'tis the chief officer in the mint, 'tis worth five or six hundred pounds per annum, and has not too much business to require more attendance then you can spare."

Newton swore out an oath to keep secret the mint's technology for making new coins, and he signed it on May 2—with that, he became the mint's new warden. In this capacity, he would oversee an annual budget of £7,500, or the equivalent of more than £700,000 today (nearly a million and a half dollars). Plus, this job would bring him to London, which was a much more interesting place to live than Cambridge. Cambridge was a small town, whereas London was a major metropolis with a population of a half million.

Leibniz failed to see the value in this move, however, and he expressed regret that the job had apparently pulled Newton away from his serious work in science and mathematics. This was true, in a way. Although Newton did not give up mathematics entirely, his creative years as a working mathematician were now long gone. In the legacy of notes, unpublished papers, and other pieces of writing that he was to pass to his heirs upon his death, are many papers and letters written after 1696 that had related to mathematics, but most of these concerned revisions of the *Principia* and were far from original, new works. He did do a considerable amount of work in lunar theory, theories related to atmospheric refraction, and the determination of a form of solid of least resistance, which were all applications that heavily depended on his mathematical ability, but these were dwarfed by the considerable literature that Newton produced on matters relating to the mint, which he threw himself into despite Montague's assurance that it was a job he could do with little effort.

Perhaps, though, Leibniz's worry about Newton's new career had little to do with the demands of the mint itself. Leibniz was no stranger to mint operations—at least in theory. He had drafted a memorandum for Ernst August years before, in which he proposed a new way of coining to take into account the fact that the Hanoverian region had some of the best silver. Leibniz had suggested introducing equivalent—as opposed to actual—weights for the coins. So a lighter coin from Hanover would be worth the same as a heavier coin from another region. This way, the value of the superior silver could be accounted for.

Leibniz may have been subtly or subconsciously referring to his own situation, pulled as he was away from more serious matters by the dreaded history that he was constantly having to work on. Perhaps he was simply expressing for Newton what he wished he could have for himself—freedom from the tedium of the history he was writing and the day-to-day petty intrigues of the court he served. Had Leibniz *his* choice, he would probably have preferred to spend his days conversing and writing on important matters. As it was, the court intrigues in Hanover at the time were enough to make a soap-opera-loving housewife blush.

The best way to illustrate that morally unhealthy setting is to illustrate it with a story, and the most intriguing one involving George Ludwig, the son of Ernst August, his wife, and his wife's lover in the 1690s. It started when George Ludwig married his cousin, Sophia Dorothea, the daughter of his uncle, George William. George Ludwig was cold and stern, in contrast to his gorgeous and affectionate wife, who was said to be attractive and well loved, the only child of her parents to survive childbirth.

Court life in Hanover in the second half of the seventeenth century was grand despite the wanton destruction wreaked in the rest of Germany during the Thirty Years' War. Ernst August was said to have stables with six hundred horses, twenty coachmen, and dozens of smiths, grooms, horse doctors, and other helpers. The halls were filled with chamberlains, ushers, pages, physicians, fencing masters, dancing masters, barbers, musicians, cooks, gentlemen of the bedchamber, and others. Entertainment was lavish, and Ernst August turned Hanover into a lavish playground celebrating his tastes. Its features included a new palace, an Italian-style opera house, and a months-long carnival.

Sophia Dorothea arrived in all innocence to marry George Ludwig on November 21, 1682. Little did she know what misery lay in store for her. Her father-in-law had a mistress in residence, the Countess Platen, who plotted against her. In the seventeenth and eighteenth centuries, fashionable men of Europe had mistresses, and not to have one was considered strange—even unmanly. However, in Hanover, the nobles' rompings tipped into incest. The countess enticed her sister to have an affair with young George. When he tired of her, the countess encouraged her own daughter (Ernst August's daughter, George's own half-sister) to become George's mistress.

Into this incestuous scene rode Count Philip Christopher von Königsmarck, a dashing young noble from a well-to-do Swedish family. He was a friend of George's brother Charles, and at a masquerade ball he met Sophia Dorothea whom, as fate would have it, he had met years before and nurtured a boyhood crush on. In 1688, he became smitten with her, and he returned a year later to settle down in Hanover, becoming a colonel in the service of the duke and settling into the welcoming arms of Sophia Dorothea, who had by then learned a thing or two about court life.

Unfortunately, Königsmarck was not a one-mistress sort of lad, and he slept with the much older Countess of Platen on the side. To her, he boasted of his affair with Sophia Dorothea. This was a very dangerous game Königsmarck was playing. His boss, Ernst August, was not a gentle and forgiving man, and Königsmarck was taking a great risk by sleeping with both his daughter-in-law and his mistress.

The duke was dangerous. He had an accomplice of his son Charles, who was involved in a plot to wrestle some inheritance from George, killed in a most heinous way in 1691. The plotter, a von Moltke, was "broken on the wheel" as they called it: His arms and legs were smashed to bits by a heavy cart wheel and then von Moltke was strapped prostrated to the wheel, which was raised on a pole and left in the sun so that he died slowly, his butt up in the air, with all the blood rushing to and swelling his broken limbs.

Königsmarck grew increasingly jealous of having to share Sophia Dorothea with George Ludwig, and when she had to spend much time in official duties during a three-month festival to mark the duke and duchess's becoming electors, he flew into a rage. In such a state, he rejected the matronly Countess Platen as a poor substitute for his young and beautiful lover. He began to blame Platen for all his troubles, including some of his financial ones, and swore he would pick a fight with the countess's son—a duel that would have been deadly for the boy because the dashing Königsmarck was a master swordman.

Rejected as a lover and threatened as a mother, Countess Platen had spies watch the lovers' every move. When Sophia Dorothea and Königsmarck decided to cut and run away together, the countess told all to Ernst August. The duke flew into a rage and had his men intercept the dashing and desperate Königsmarck before he could rendezvous with his beautiful daughter-in-law. In the melee that ensued, Königsmarck was cut down and, as he died, the Countess Platen, who had been waiting in the shadows watching throughout the ambush, stood over him. It is said that, with his last breath, he cursed and spit at her, and that she dug her heel into his mouth and twisted aside his curses.

Such were the pleasantries of the intrigues at Hanover during the 1690s.

The official account of the Swede's disappearance simply stated that Königsmarck had wandered off on that night, never to be seen again. But the damage was done. The lovers' letters had been found and the scandal took on a very public persona throughout Europe. In a stab at damage control, Sophia Dorothea was put on trial, and a divorce was granted on December 28, 1694. George Ludwig was now free to remarry. Sophia Dorothea, on the other hand, became a prisoner at a nearby fortress, and her children were taken from her. She lived thirty-two more years alone and forgotten—abandoned by her unfaithful husband, bereft of her murdered lover, and missed by her children.

In the years to come, George Ludwig grew to dislike his children— especially his son, who greatly resembled Sophia Dorothea. George was indifferent when his grandchildren were born, and by then was so nasty toward his own offspring that, a few years after he became king of England, he ordered his grandchildren, ages five, seven, and nine years old, forcibly removed from the children's parents. His orders even went so far that his newborn grandson was ripped from his mother's arms and died a few weeks later—possibly as a result of the heavy-handed act.

In this interesting yet empty dramatic scene that was Hanover toward the end of the seventeenth century, Leibniz languished with little if any intellectual company. He confided to Thomas Burnet in 1695 that he simply had hardly anyone to talk to and that, if it were not for his discussions with the aging Queen Sophie, Ernst August's wife, he would have almost none. He had to rely upon his correspondence with people like Burnet for intellectual company, and, from his point of view, Leibniz could not wish that anyone else, especially a mathematician as brilliant and seemingly fragile as Newton, be mired in such tedious governmental intrigues. In 1696, the same year that Newton began working at the mint, a curious thing happened to Leibniz—he nearly married. While he was in Frankfurt at one point in his travels, some of his friends suggested he pursue a rich young spinster, and apparently he did make certain overtures, but it was probably closer to a legal negotiation than an intimate wooing, and nothing came of it. In the end the lady took her time to consider his proposal, and he lost interest. He was fifty years old at the time, and a lifelong bachelor. For my part, I can't help but wish I knew more about her.

9

Newton's Apes

Know yee that wee for divers good causes and considerations . . . do give and grant unto Our trusty and Well beloved Subject Isaac Newton Esqr. the office of Master and Worker of all our Moneys both Gold and Silver within our Mint in our Tower of London and elsewhere in our Kingdom of England."

—William III, king of England. *Newton's Appointment as Master of the Mint*, February 3, 1700

Newton had a house at the Tower of London, in which he lived for the briefest time when he moved to the capital. By the end of the seventeenth century, when he took up residence there, the Tower was already an ancient site that was steeped in history and intrigue. Newton lived just a few short steps from where Anne of a thousand days and many other famous prisoners had been executed.

Today, Mint Street boasts a narrow row of rather unassuming black brick houses on the westernmost side of the Tower complex; they have been converted into private homes which a docent told me are inhabited by people who work at the famous tourist attraction.

Newton did not live there for very long. The din of all the coin clapping was so bad that he soon found a house in another part of town. But sleeping far away from the noise did not mean his attention was elsewhere. His mint was a government office in crisis. By decree, the mint had been charged with recoining the silver currency of the realm—something that became necessary because the old coins had smooth edges and could be easily "clipped."

Clipping occurred when dishonest types would snip a little sliver of metal from the edge of a coin. If they did this to enough coins, they could melt the pieces into a bar of silver bullion; and because one of the intricacies of the mint was that it allowed people to exchange silver bullion for silver coin, the clippers could trade this bar for new money.

If coin clipping was a chronic problem, counterfeiting was an acute one. For most of the seventeenth century, England's silver coins were struck by hand—a grueling piecework task involving sweat and hammered silver dyes. This method had been abandoned thirty years before Newton came to the mint; instead, the coining had become more of an industrial process whereby silver was melted in massive iron pots over coal fires and coins hammered out by machines specially designed for the task. But some of the old coins were still in circulation and, as long as they were, counterfeiters could create their own dyes and hammer out fakes made from lesser alloys.

Recoining the currency was the solution to these issues because a new invention allowed the new coins to be edged (given a distinctive rim) in a machine when they were made, which prevented them from being clipped undetected and also made counterfeiting much more difficult. So the mint, which was located on Mint Street between what was once the inner and outer walls of the Tower of London, began churning out new coins at the end of the seventeenth century. Some three hundred workers and fifty horses turned the nine minting mills from 4:00 a.m. to midnight every day, and they cranked out 100,000 pounds of coins per week. It was the largest recoinage program in England's history—and was not going well.

When the project began, it had a number of problems. For one, there was not enough income to pay for the expense of the operation. The mint was funded by a tax on imported liquor, which was not enough to support such massive recoinings, and as a solution the government imposed a new tax on windows in the city of London. Supposedly, some of the old "blind" windowless buildings in the city survived from that time even to the twentieth century.

Moreover, the new coins were still not a guarantee against counterfeiting, and there were lots of counterfeiters clever enough to beat the new system. New silver coins were cast from 92.5 percent pure silver and 7.5 percent copper. A counterfeiter could simply buy new coin, mix it with copper or lower-grade silver, and beat out counterfeit coins, and then exchange these for more new coins. This was so easily done and so rampant that by the time Newton became warden of the mint, he estimated that a fifth of the coins the mint took in were fakes.

Newton got to know the ins and outs of counterfeiting and clipping quite well when he took his first job as warden, one of the mint's three chief officers. As warden, he was the king's representative at the mint, a post that had once been ostensibly its top official. He managed the mint's finances and supervised its other officials, but really the power of the mint resided with the master of the mint, who was sort of the head contractor. The master's contract was simple. For every pound of silver he minted, he was allowed a certain percentage as commission, and with this he subcontracted the work and took his profit. By the time Newton was hired, the functions of the warden's office had been reduced, and the master of the mint had assumed a great deal of power and no longer played second fiddle to the warden.

Basically, the warden was responsible almost exclusively for police and legal work. Newton's first duty was ferreting out counterfeiters and clippers, and prosecuting them—work that held little appeal to him but which he excelled at. While the prosecution of the coiners and clippers was a duty that had been a part of the warden's job for decades, Newton's predecessors had left it to their clerks to carry out. Newton did it himself but is said to have been so disgusted by the work that after a while asked the treasury to relieve him of it. "'Tis the business of an Attorney and belongs properly to the King's Attorney and Solicitor General," he wrote. "I humbly pray that it may not be imposed on me any longer."

This is not to say that he slacked off on the job. He took to his prosecutions with the same singular zeal that he applied to most things in life, personally taking extensive depositions from the accused counterfeiters and their lawyers, and writing something like a case-book to guide his work. He even bought a new suit for the task. He paid a significant sum of his own money to be made justice of the peace in several counties so that he could prosecute counterfeiters far and wide.

If there was any one criminal on whom Newton would sharpen his prosecutorial teeth more than any other, it was the notorious counterfeiter William Challoner. Challoner was a thief and a flim-flam man of great skill and even greater bravado. A few years before Newton became warden, he had managed to collect a handsome reward through a shrewd backstabbing con involving a bounty offered by the British government for information leading to the capture of a pamphleteer who was spreading propaganda against the king. Challoner found one of the offending pamphlets, paid to have it reprinted, and turned in those printers for the reward money.

In early 1696, when Newton arrived at the mint, Challoner had an even bolder con in the works. A year earlier, he had written a pamphlet advocating a reduction of weight of silver coin to match the older, clipped coins. Presumably the reason for this was that Challoner was one of the best counterfeiters in Britain at that time, and a reduction of weight would mean more profits for him, since he could use less silver in his counterfeit coins. He approached Parliament and various members of the British government to decry the incompetence and corruption of the mint. Offering his services, Challoner claimed that he had invented a way of making counterfeit-proof coins. He tried to convince the government that he could modernize the coin-pressing machines at the mint, provided he personally supervised their operation.

A parliamentary committee that heard these offers asked Newton a few months later to give Challoner access to the mint machines. But Newton refused. Some of these machines were top secret. Newton himself had to swear out an oath not to reveal the workings of the mint operations when he assumed his post. He seems to have seen through Challoner's ruse, and was not one to be trifled with in these matters. Newton had Challoner clapped in irons and placed under arrest. Many months later, in 1699, he had prosecuted the notorious counterfeiter so successfully that the rogue was put to death for his crimes.

For this and other displays of competence, Newton was rewarded with a promotion to master of the mint in 1699, when the master he served under, Thomas Neale, died. Newton was appointed in his place the day after Christmas. His appointment letter, in the name of King William III, granted "unto the said Isaac Newton all edifices, buildings, Gardens, and other fees, allowances, proffitts, privileges, franchises and immunities belonging to the aforesaid Office."

Soon after this appointment, the spheres of Newton's government and Leibniz's would become inexorably linked. England's King William died in 1702; and his co-regent wife, Mary, had had the bad fortune of fatally falling victim to smallpox about ten years before (and some hundred years before Jenner's method of vaccination began the long process that would eventually lead to the eradication of the disease from the world in the 1970s). William and Mary left no heirs to the throne, and this set up something of a crisis in the years before the king died, when the British Parliament scrambled to find some solution to ensure a Protestant succession.

Next in line was Princess Anne, who was safely Protestant despite the fact that she was the daughter of the deposed King James II. In 1702, after William died, she became the ruler of Britain. Queen Anne was a stout, squat woman. In a 1705 portrait of her by the artist Michael Dahl that hangs in the National Portrait Gallery in London, she is standing impressively in a regal gold dress and royal blue furry wrap, wearing impossibly large diamonds, one hand placed on the crown jewels. Having herself portrayed standing may have been an important statement because, when she was crowned in April 1702, she was obese, sick, and stricken with gout so severe that she could not stand or walk. She had to be carried on the backs of the yeomen of the guard into Westminster Abbey, where the crown was placed on her head.

Hers was not an easy rule. She inherited a war, which began in 1702 when a new alliance—the "Grand Alliance" of Denmark, Prussia, Hanover, the Palatinate, England, and Holland—was formed against France; and her reign also coincided in almost its entirety with the war of Spanish succession, which was not concluded until the Treaty of Ultrecht the year before Anne died in 1713. The real tragedy of Anne's life, though, was that she was unable to raise a child, despite half a lifetime of trying. Queen Anne had eighteen pregnancies and seven children, but the last one died before she even took the throne.

Even before she became Queen, however, the British parliament passed in 1701 the *Act of Settlement*, which explicitly named the descendants of Leibniz's good friend Queen Sophia as being in line for the British crown. This meant that Sophia's son George Ludwig was slated to become king of England after Anne died. George was in an odd position to be in line for the throne. Proper succession would have had Queen Anne's brother, James (the would-be James III), become the king. Instead, because of the *Act of Settlement*, the crown passed to their distant cousin, whose sole claim was that his great-grandfather on his mother's side was King James I. James's daughter Elizabeth Stuart had married Frederick, an elector in Germany, and they had had one child. She was the Sophia who wound up marrying Ernst August, the Duke of Hanover. Sophia and Ernst had six children, one of who was George Ludwig; when Ernst August died in 1698, he was succeeded as duke by George. The cause for George's ascension to the English throne would not have looked good were he subject to rules of inheritance that were not based on the fear of another Catholic king.

After George Ludwig became elector of Hanover, he changed the good-natured mood of the court. One witness to this change, the Duchess of Orleans, described it in a letter: "It is no wonder that pleasure is no longer to be seen in Hanover, for this Elector is so cold he turns everything to ice."

There was some friction between Queen Anne and the duke in Hanover years before. George Ludwig traveled to England and was presented to Anne as a potential husband, but nothing came of the meeting. Rumors at the time attributed this to the fact that the fit and fierce young duke was not attracted to the squat, square Anne.

There was also friction between George Ludwig and Leibniz after the former became duke. George was probably willing to put up with more from Leibniz than he would have from one of his other courtiers, since Leibniz was a living legend—an illustrious thinker who was a leading intellect. Besides, Leibniz did provide valuable advice and faithful service. However, George never really took Leibniz into his confidence, and his respect often took a mocking tone. He once referred to Leibniz, for instance, as his "living dictionary," and complained of his frequent absences and inability to finish the Brunswick history.

Leibniz took these things in stride. He was, indeed, a living legend by the time George Ludwig became duke in 1698. His record of service to the House of Hanover was solid and his list of accomplishments, quite aside from his work for the two previous dukes, George Ludwig's father and his uncle, was immense. Despite the fact that he had not finished the history on which he had already been working nearly ten years, Leibniz assumed his court position at the end of the seventeenth century was as secure as his being the grandmaster champion of calculus—unassailable.

Hardly!

Out of nowhere, Fatio de Duiller popped back into the picture, when he was moved to champion Newton in 1699 after writing an article, "A Two-Fold Geometrical Investigation of the Line of Briefest Descent," which made the startling public accusation that not only was Newton the first to discover calculus, but that Leibniz had actually stolen it from Fatio's mentor and friend.

"The celebrated Leibniz may perhaps inquire how I became acquainted with the calculus which I may use," Fatio wrote. "I recognize that Newton was the first and by many years the most senior inventor of calculus, being driven thereto by the factual evidence on this point. As to whether Leibniz, its second inventor, borrowed anything from him, I prefer to let those judge who have seen Newton's letters and other manuscript papers, not myself."

Why did Fatio suddenly jump in where there didn't seem to be any ongoing dispute, and champion Newton after they had, for the previous several years, drifted apart? One possibility is that perhaps he was seeking to renew his friendship with Newton. But equally compelling as an explanation is that he may have been driven by feelings of resentment toward Leibniz.

Fatio had his own personal history with Leibniz, and he disliked the German immensely. Just as Leibniz had done a decade earlier, Fatio had made a connection with Huygens. Fatio was younger than Leibniz had been when he was in the same position, and Huygens was much older, but Fatio nevertheless saw himself as something of a peer of Leibniz because they were now both disciples of Huygens. For his part, Huygens wanted to promote an exchange between Fatio and Leibniz because he thought that it would be productive, and so Fatio wrote to Leibniz on several occasions, asking him to share his mathematical tools and techniques. Leibniz refused to become involved, failing to see what he stood to gain from the exchange. He still had the greatest respect for his old mentor but apparently not enough for Huygens's new protégé.

Perhaps it was this earlier snub that led Fatio to take up Newton's cause in 1699 and accuse Leibniz of plagiarism. Or perhaps he felt slighted by Leibniz because of Bernoulli's challenge problem, which Fatio also solved but didn't get it in on time. Fatio was offended when he read what Leibniz wrote about this problem—the gloating boasts about how only Newton and Leibniz's exclusive clique of followers could solve it. Fatio saw this as a direct snub and was driven, therefore, to counter by throwing sand in Leibniz's face in the form of a major accusation of plagiarism.

"Neither the silence of the more modest Newton nor the eager zeal of Leibniz in ubiquitously attributing the invention of this calculus to himself will impose on any who have perused those documents which I myself have examined," Fatio wrote in his paper. There is no question that he was uniquely positioned to make such a spirited defense of Newton and attack of Leibniz. Fatio was one of the few people in Europe who were sufficiently versed in calculus to understand the documents, plus he had been given more access to Newton's inner sanctum, with all its rich papers, than nearly any other human in the Englishman's lifetime.

Nevertheless, Fatio's attack was ill timed. While he did not outrightly accuse Leibniz of plagiarism, he definitely implied it. A substantive case against Leibniz would eventually be made, and in time many others would also take up the accusation and attack Leibniz on Newton's behalf much as Fatio did in 1699. But that would have to wait until later. In 1699, Newton was just a few years into his tenure at England's mint, and overseeing this institution consumed much of his time and attention. He offered no help to Fatio.

Operating alone, Fatio was very far out of his league. Leibniz was, after all, hailed in Europe as the foremost mathematician of his day—a position that Leibniz himself was ever vigorous in asserting. In England, too, his reputation was stellar, and he was a long-time member of the Royal Society. And the foremost mathematician of his day was furious. He showed amazing restraint by not losing his cool; instead, he replied directly in the pages of the *Acta Eruditorum*.

In May 1700, Leibniz published his response to Fatio's accusation, defending his position vigorously and dismissing the young man as having been perverted by a thirst for recognition. In an almost psychoanalytical attack, he wrote, "Mistrust is a feeling of hostility." And he followed this statement with the eloquent zeal of a brilliant lawyer: "We can readily conceal under a zeal for justice sentiments which, plainly acknowledged, would disgust us. In truth, the more I understand the defects of the human mind, the less I grow angry at any aspect of human behavior."

In his article, he defended himself by implying that Fatio did not have the support of even Newton in his accusation. For Leibniz, even though his relationship with Newton had always more or less been at a great distance, theirs had all the outward signs of one of mutual respect and the highest admiration. Newton's silence on the issue was deafening to Leibniz's ears. "At least the excellent man appeared, in several conversations with friends of mine," wrote Leibniz, referring to Newton, "to manifest a kind disposition towards me, and made to them no complaints, so far as I know. In public, also, he has spoken of me in terms which it would be most unjust to find fault with. I, too, have acknowledged his great service on appropriate occasions."

Leibniz was willing to give Newton his due as a mathematician on more than one occasion, and this was certainly one of them. At this point, the German still had no quarrel with his British rival, and he did not waste the opportunity to give adequate if not overabundant praise to Newton, setting themselves out as mathematical equals. But he maintained that theirs were parallel greatnesses—that he had gleaned little of Newton's original discoveries from their exchange of letters. He claimed that he had no idea how advanced Newton's mathematics were until he read the *Principia*, but that it was not until the 1690s that he realized that Newton's methods were "a calculus so similar" to his own. In his article, Leibniz pointed out that the Englishman, in the *Principia*, established that they respectively invented their mathematical methods independently: "As in his *Principia* he has also explicitly and publicly testified, that neither of us is indebted, for the geometrical discoveries made common by us both, to any light kindled by the other, but to his own meditations."

Leibniz also explicitly stated his innocence in the matter. "When I published my elements of the differential calculus, in 1684, I knew nothing of his discoveries in this department, except what he himself had told me in one of his letters, wherein he stated that he could draw tangents. ..."The drawing of tangents that Leibniz drew attention to in this paper (an operation that is greatly simplified by the use of calculus) was hardly unique to Newton. Likewise, elsewhere Leibniz explicitly pointed out that nobody knew better than Newton how their discoveries were truly independent "without either receiving any enlightenment from the other."

Leibniz did not merely write a rebuke of Fatio's paper in his own favorite journal. For good measure, he also reviewed his own letter anonymously, giving it a favorable review of course. In addition, he sought vindication by complaining formally in a letter that was presented to the Royal Society on January 31, 1700. Without Newton's backing, Fatio was easily shot down by Leibniz, and the mathematicians of note in those days backed Leibniz.

John Wallis, for instance, was said to have been most distressed by the accusations and sympathetic toward Leibniz. He assured him that Fatio's attack had not been sanctioned by the Royal Society and that Leibniz's reputation was safe. And Newton? . . . Newton remained silent on the matter.

The dispute could have ended here, and the calculus wars could have fizzled out with Leibniz, the victor of sorts, allowing that Newton was his equal in original discovery, demonstrating that Fatio was out of line, and going on with his business. To Leibniz, this was a simple matter that had been simply resolved. In his worldview, the invention of calculus belonged to him more than it did to Newton. Had they not discovered it independently, and had not Leibniz published his work first? "When I published the elements of my calculus in 1684," Leibniz wrote, "there was assuredly nothing known to me of Newton's discoveries in this area, beyond what he had formerly signified to me by letter."The material he was referring to, he added, was not calculus but rather some preliminary methods.

Moreover, Leibniz had published calculus in a journal that was then being circulated among the top mathematicians in Europe. His methods were long established and well known throughout the Continent, not squirreled away as if some guilty secret. And, most important, had he not invented the notation of calculus that allowed its further development? In 1700, his calculus was successful in various applications used by others with Leibniz's blessing, and the fact that it continued to be developed was strong testimony to Leibniz's methods. Newton, on the other hand, did nothing to publish his version of calculus until he was a relatively old man, and he seemed less interested in promoting his fluxions and fluents than in securing the rights of their invention for himself; moreover, his notation was inferior to Leibniz's.

Solidifying his mathematical reputation, Leibniz published another paper in 1701 under the French title, "Essay d'une nouvelle science des nombres." The essay was in honor of his being made a member of the French Academy of Sciences, and it described a new science of numbers called binary mathematics, which he had developed in 1679. Binary (literally, "two numerals") is a system whereby all values are represented as sequences of only two digits—one and zero. Leibniz thought that binary numbers would reveal properties of ordinary numbers that would not otherwise be apparent, and in fact binary numbers, as established by Leibniz, became the basis of electronic circuitry.

Following Leibniz's well-presented series of rebuttals, Fatio did not fare so well. In 1704, he was the secretary to a group of fanatics called the Camisard prophets—a sort of doomsday cult from France who were obsessed with the imminent fulfillment of prophecy from the Bible's revelations and who claimed that they could raise the dead. The group was ostracized for their beliefs, and Fatio himself was pilloried at Charing Cross on December 2, 1707. His head and hands were stuck through the holes of the wooden frame of the pillory, and a hat was placed on his head that read, "Nicolas Fatio convicted for abetting Elias Moner in his wicked and counterfeit prophesies and causing them to be printed and published to terrify the queen's people."

Interestingly, Leibniz never seemed personally vindictive toward

Fatio even after the accusations came out. Several times after the events of 1700, he addressed kind words about Fatio in writing to his friend Thomas Burnet. And when Fatio was pilloried in 1708, Leibniz wrote of how appalled he was by the treatment though also at how Fatio, "a man excellent in mathematics," could have been involved with the Camisard prophets.

The dispute with Fatio portended a different sort of doomsday for Leibniz. Fatio's attack was isolated and little came of it. But it was a signal of what was to come.

The next time the fires flared up was when they were stoked by a minor character named George Cheyne, whose main claim to fame other than his role in the calculus wars seems to be his strange new theory of fevers, which he based on Newtonian physics.

CHEYNE WAS SCOTTISH by birth but had settled in London around the turn of the eighteenth century as one of a growing group of Newtonians. In an unauthorized tribute to his new master, Cheyne wrote a book he called *On the Inverse Method of Fluxions*, in which he attempted to explain Newtonian calculus to the world.

It was an inferior, unimportant book by a man who would probably have been completely forgotten had it not been for the fact that so little had ever appeared in print on the methods of calculus that it could not have gone unnoticed. And indeed many people noticed it—not the least of them Newton.

When Cheyne's book was published, Newton was becoming more and more important as a figure in England. Robert Hooke died in March 1703, and this freed Newton of his longtime nemesis, who had been a cantankerous gadfly to him at times. Even at the end of his life, Hooke was still menacing Newton with his public accusations. On August 16, 1699, for example, when Newton appeared before the Royal Society to present a sextant he had just invented, Hooke, always unimpressed, responded by claiming that he himself had invented the sextant thirty years before.

Shortly thereafter, on November 30, 1703, Newton was elected president of the Royal Society. This was not the only satisfaction Newton enjoyed at the turn of the eighteenth century. On April 16, 1705, he was awarded the ultimate recognition of knighthood by Queen Anne.

Now, as a knight and Royal Society president, Newton was finally about to throw off his long silence and assert his priority in the invention of calculus when he published *Opticks* in 1704. Cheyne's book was part of the inspiration for this, because Cheyne got Newton's calculus wrong enough that Newton wanted to get his own written material out there, which he did in the appendix section of *Opticks*, "On the Quadrature of Curves."

This led directly to a confrontation with Leibniz, because after he became aware of *Opticks*, Leibniz of course leapt to publish an anonymous review of Newton's mathematical appendix. In his review, he wrote, "Instead of the differences of Leibniz, Newton applies and has always applied fluxions . . . as also Honoratus Fabrius, in his Synopsi Geometrica, substituted progressive motion in the place of indivisibles of Calvalieri."

What did he mean by this? It means almost nothing to modern readers, the names Honoratus Fabrius and Calvalieri being so obscure that the offending statement is completely vague—even innocuous. But to a mathematician as brilliant as Newton, who was well versed in the mathematical discoveries and controversies of his day, the meaning was instantly clear. Fabrius had borrowed the work of Calvalieri, and by comparing Newton to the former, Leibniz may have been subtly implying that Newton borrowed calculus from him. This would really be too much for Newton to endure when he found out about it.

However, it would take a few years before Newton did find out, and those years would be the last that Newton and Leibniz would spend, that weren't clouded by the full-blown calculus wars, which would explode after 1708, when one of Newton's supporters would attack Leibniz.

Meanwhile, the years between 1705 and 1708 were not the happiest of Leibniz's life because of the loss of a good friend. For years he had been close to the women of the German courts. He was a perfect companion for the ladies of the court, really, since he could speak wonderfully and was well informed on a dozen topics of timeless importance and probably twice that many topics of contemporary or trivial interest.

Particularly endeared to Leibniz was Sophie Charlotte, the daughter of Queen Sophia and Ernst August, who had an extraordinary affection for the older philospopher. She once expressed this in the over-the-top superlative praise that was the fashion of that time. "Think not that I prefer this greatness and these crowns, about which they make such a bustle here," she wrote to Leibniz, "to the conversations on philosophy we have had together."

Sophie Charlotte was an important royal in Europe. Ernst August had married her to Prince Frederick of Brandenburg when she was a teenager. Sophie Charlotte was a lovely girl, beautiful, rich, intelligent, and destined for greatness. Her husband became the elector Frederick III a few years after they were married, and, a while after that, in 1701, Frederick and Sophie Charlotte became king and queen of Prussia. Their grandson was Frederick the Great.

Sophie Charlotte had been tutored by Leibniz and as queen carried on an extensive correspondence with him—on metaphysics, history, literature, and just about everything. She was apparently so clever that she sometimes complained that Leibniz oversimplified things in his discussions with her. Supposedly, according to Frederick the Great, when she complained to Leibniz about this, he said that it was a reflection of her brilliance more than his condescending attitude. "It is not possible to satisfy you," Leibniz supposedly said. "You desire to know the wherefore of the wherefore."

Her death in 1705 was such a shock to Leibniz that certain ambassadors and dignitaries in Berlin paid their respects to him as though he were the closest surviving family member. He later wrote one of his most famous books, *Theodicy*, based on conversations he had with Sophie Charlotte and on writings he did for her in French that had been based on the same conversations, as a sort of memorial to her. It addressed questions of church doctrines that he had first wrestled with in his efforts toward church reunification at the end of the seventeenth century. The book was a very influential work after it was published in 1710, especially in Germany, and is one of the most important primary texts for Leibniz scholars today because it expresses Leibniz's philosophy. *Theodicy* was published anonymously in 1710, because Leibniz did not want his name to appear on a theological work.

Publishing anonymously was very common in the seventeenth century, and Leibniz had already found it a convenient way to express his mathematical opinions at certain times. This sort of anonymity complicated things immensely in the years to come because the communications that passed back and forth between Newton and Leibniz was often marked by subterfuge. Both men relied on their supporters to make their arguments and attacks for them. Leibniz had the advantage on paper, since he had a few key supporters in Europe who were themselves brilliant mathematicians. But curiously, Newton had the real advantage—perhaps because he was without equal among his supporters, as would be demonstrated by one of Newton's key followers, a young Oxford professor named John Keill, who made prosecuting Newton's case against Leibniz his own personal crusade.

Keill was a Scot who had followed his teacher, David Gregory, to Oxford in 1694. Though a very minor character on the stage of science, he turned out to be a major player in the calculus wars. Much as Fatio had done, Keill sought to go beyond securing Newton's due credit as co-inventor. He wanted to secure all the credit and accompanying fame for Newton and Newton alone. In order to do this, Keill had to show that Leibniz had stolen calculus from Newton. Eventually receiving a great deal of help from Newton, Keill succeeded in issuing a serious challenge to Leibniz's credibility.

Newton's second "ape" after Fatio, Keill went on the offensive in

1708 and began to accuse Leibniz of plagiarism. He published a paper in the Royal Society's *Philosophical Transactions* in late 1708, though it was not printed until 1710. Keill's paper was a minor little pastiche on physics that inexplicably contained a major accusation. "The Laws of Centripetal Forces," as it was called, is more noteworthy for what it said about the calculus dispute than for what it said about centripetal forces. In it, Keill wrote that Leibniz's calculus was "the same arithmetic" as Newton's fluxions, and he called Newton "beyond all doubt" the first inventor of fluxions: "The same calculus was afterward published by Leibniz, the name and mode of notation being changed."

Keill's claim was carefully crafted to be a blunt but indirect accusation of plagiarism against Leibniz. Nobody could dispute that Leibniz had published first. So Keill chose the next best thing. He said that Newton had invented calculus prior to Leibniz and Leibniz did not follow Newton merely in time but also in design. Moreover, Keill modified his attack in such a way as to state that he was not accusing Leibniz of plagiarism, while at the same time suggesting that plagiarism is exactly what the German had done. Even though Newton didn't write down his ideas and share them with his contemporaries through publication, he had nevertheless shared them with Leibniz. Keill stated that Leibniz could have gotten everything he needed to develop calculus from the two letters that Newton had sent to him way back in 1676. They contained, Keill said, what was "sufficiently intelligible to an acute mind."

It was a clever approach, really. The case became one that was winnable for Keill—and ultimately for Newton—because they were not trying to prove historically that Leibniz had stolen anything all those years ago, but rather that he merely *could* have. And by coupling this argument with the even sounder evidence that Newton had developed his methods of calculus prior to Leibniz's version, it was enough to make the case that the English mathematician was the true and sole inventor of calculus.

Such a strong challenge had not been made since the ill-fated attempt by Fatio to win credit for Newton nearly a decade earlier.

But unlike Fatio's arguments, which fell apart like a house of cards under a single wave of Leibniz's hand, Keill's attack was much more dangerous. It was a deliberate provocation that Leibniz had to answer—a bear trap covered with twigs and leaves.

The winter of 1709 was a terrible and miserable time in Europe. It coincided with military disaster for the French and a terrible famine in Europe, as unusually harsh conditions visited down upon the populace. And another war, several decades in the making, was finally about to explode.

10

The Burden of Proof

"Justice is a social virtue, or a virtue which preserves society."

—Leibniz, *On Natural Law*

Leibniz was dumbfounded and incensed upon hearing of Keill's accusations. He assumed that Keill had erred because of some rash conclusion that he made, and resented that a man whom he did not regard as one of his legitimate peers was making such accusations in the first place. Who did Keill think he was? To obtain satisfaction, Leibniz would turn to the venerable Royal Society, of which he was a longtime member. This was the same course of action he had followed when Fatio had made his unsupported attack, and Leibniz had been vindicated then, therefore he expected to be so again—not just because it was exactly the same situation but because he knew that he was right. He had not stolen anything from Newton, and he was confident the intelligent members of the Royal

Society would see things his way. Leibniz was a great believer in intellectual societies, after all.

Scientific societies were a big part of his life, as they were for many of the scientists in the seventeenth and eighteenth centuries. In his lifetime, Leibniz had seen how the academies played an important role both in the collection and communication of as well as in the conducting of experiments. The French Académie des Sciences, for instance, sponsored major projects, such as an accurate mapping of the French Empire through South America, Africa, the West Indies.

And Leibniz was especially fond of these scientific societies because he saw the greatest of possibilities for them. The existing societies in Paris and London, venerable institutions with an august membership, were but trivial gentlemen's clubs compared to what Leibniz envisioned. He had an almost unquenchable enthusiasm for the possibilities of scientific societies because they fit into his grand vision of a more perfect world, and he even had attempted to found such a society of sciences in Berlin.

In 1697, Leibniz found out from diplomat Johann Jakob Chuno that Sophie Charlotte wanted to build an observatory in Berlin, and he immediately sent her a letter saying that she should expand her plans and turn it into a scientific academy. Leibniz's designs for the Berlin Society of Sciences were complicated by the fact that there were cool relations between Berlin and Hanover. Besides that, from George Ludwig's point of view, the writing of the history of the House of Brunswick was Leibniz's main task, unacceptably overdue.

At first, George Ludwig forbade Leibniz even to go to Berlin, but eventually the duke relented, and finally, in 1700, George allowed Leibniz to make the trip—but only after the elector in Berlin had personally requested Leibniz's presence. The society was successfully launched with the support of Sophie Charlotte and Frederick III, who liked the idea that he would be seen as a patron of intellectual pursuits; and Leibniz was to be appointed the society's first president. Frederick the Great would later say that Leibniz was a society of sciences all by himself.

In a way, this was nothing new for Germany. Groups that met and discussed philosophy, physics, mathematics, astronomy, or any number of other subjects on a regular basis were probably quite common. Leibniz had belonged to one at the University of Jena, while for one semester he was working toward his doctorate in law. There, a group of professors and students met once a week to discuss new and old books. He had joined a similar one at the University of Leipzig.

But these groups were nothing compared to institutions like the Royal Society or the French Académie des Sciences. What Leibniz had envisioned for the Berlin society was even grander than its English and French counterparts. "The labors of such a society should not be directed merely to the gratification of a scientific curiosity and the performance of fruitless experiments, or simply to the discovery of useful truths, without any application of the same; but the uses of science should be pointed out, even at the outset, and such inventions be made as would redound to the honor of the originator and the benefit of the public," he wrote. "The aim of the society, accordingly, should be to improve not only the arts and sciences, but also agriculture, manufacture, commerce, and, in a word, whatever is useful in the support of life."

His vision for his scientific society was something akin to the modern think tank, but perhaps with a lot more power. Leibniz thought his society should not simply advise, study, and report on the issues of the day but that it also should establish policies, practices, and progressive approaches to improving life. He desired it to not solely to be focused on science either, but to expand its interests to include history, art, and commerce.

Leibniz had harbored this vision for years. His scheme for draining the mines in the Harz Mountains had been predicated on the notion that it could fund such a society. From an earlier experience, when he had still been in the service of the elector of Mainz and Boineburg, he had learned the value of not proposing too much, after the elector had rejected his far-reaching plans as being too ambitious and expensive. These plans, incidentally, had called for changing everything from standard units of measurement to the church's role in education, and sought to deliver a lot of the decision-making power into his proposed academy's hands.

In 1700, Leibniz had learned to rein in his plans a lot more, but his schemes were still grand, of course. His Berlin society was to have an observatory, laboratory space, hospitals, libraries, a press, and museums. He did not underestimate how much money it would take to achieve his goals; precisely because he was keenly aware of the financial needs of such an enterprise, this forced him to come up with an inordinate number of schemes to finance it.

To fund the academy, Leibniz unleashed a torrent of creative ideas. He suggested asking for donations from the church, creating a lottery, and instituting a number of new taxes, including a tax on wine, a small income tax increase, and a tax on foreign travel and paper. He wanted to obtain monopolies on the production of new calendars and almanacs, on the production of fire engines, and on the production of mulberry trees, which were used in the cultivation of silkworms.

In fact, Leibniz was so keen on mulberry trees that he tried for years to get them to grow. However, this was a failure because the silkworms did not thrive in the Germanic climate. The mulberry plantations were eventually abandoned and fell into ruin.

Like his mulberry trees, Leibniz's grand vision fell into ruins as well. The problem was that the academy was in Berlin and he was in Hanover, and although he now had a legitimate reason to travel, he nevertheless had to obtain permission from George Ludwig each time he wished to do so. The duke, of course, had no interest in allowing Leibniz to spend long periods of time away from Hanover—not while the history needed his attention.

Leibniz's situation was further complicated by the fact that relations between the courts at Hanover and Berlin were strained; this even led to his being accused of being a spy when he was in Berlin. The effect his absences had was to reduce Leibniz's influence at the academy. He may have held the official title of president, but for most of the time in those early years of the academy he was out of sight and out of mind.

The two academy members who really had the power were a pair of characters known as the Jablonski brothers. One was secretary and the other the acting president. They eventually stopped consulting Leibniz on the appointment of new members, and added the ultimate insult by electing a Baron von Printzen as director of the academy in 1710. When the academy was officially inaugurated on January 19, 1711, Leibniz was not there, and in April 1715, his salary was abruptly halved. The final insult was that, when Leibniz died a year and a half later, the academy did nothing to mourn the passing of its creator.

Nevertheless, even if his own Berlin society had not turned out the way he envisioned in 1711, when he prepared to respond to Keill's attack, Leibniz was still a great believer in scientific societies in general, and he had a great deal of respect for the Royal Society and felt they would justly decide his case if he put it before them.

For Newton, the only scientific society that really mattered *was* the Royal Society of London. When he became its president on November 30, 1703, the society had changed somewhat since its glory days in the 1670s, when Newton had been elected among its hundreds of members and it oversaw many important experiments. These issues were a faint memory in 1703. New membership was stagnant and total membership had declined.

The types of discussions and experiments at the Royal Society had become the subject of ridicule. Jonathan Swift satirized the Royal Society in *Gulliver's Travels* by describing scientists who wanted to extract sunshine from cucumbers. England's king was reportedly amused by the attempt by one Royal Society member to weigh air, and some of the society's genuine discussions of the medicinal properties of common or uncommon substances are equally comical. In 1699, one Royal Society fellow, a Mr. Van de Bemde, remarked how cow piss "drank to about a pint" will cause a person to either purge or vomit "with great ease."

But Newton brought renewed vigor to the Royal Society, and for the next twenty years, he ran it as a CEO would manage a personally financed startup. Newton presided over almost every meeting the society had for the next couple of decades, including the smaller meetings of the society's council. His tenure was unusual. Every president prior to Isaac Newton had only served a few years at the most, and some had administrations so short they could almost be called "acting" presidents. Samuel Pepys, for instance, was president for exactly two years in 1684–1686, and Christopher Wren was also in office for two years, starting in 1680.

It's no exaggeration to say that when Leibniz made his appeal to the Royal Society, he was really making his appeal to Newton himself. Newton *was* the Royal Society in those days.

———

THE YEAR THAT the calculus wars exploded into a full-blown battle, 1711, was a time of increasing accomplishments for Newton, in terms of publishing. A few years before, in 1707, William Whiston had published in Latin Newton's Cambridge lectures on algebra, *Arithmetica universalis*, which the mathematician gave in accordance with the requirements of the Lucasian chair he held. As far back as 1672, Newton had begun compiling notes for these lectures. In 1712, the text for this would be translated and published in London in English. Meanwhile, Newton's *De Analysi* was published under the editorship of William Jones. This book basically demonstrated some of the results obtained with Newton's calculus, but with no formal treatment or notation. Jones had purchased John Collins's library several years after Collins had died, and there among the books and papers he found Newton's text from so many years before. Jones sought out Newton for permission to publish the book, which was granted, and he brought out the edition in 1711.

Leibniz probably could not have cared less about these publications. They were taken from material that was badly out of date, having been written decades before. He was more interested in the outrageous accusation published by Keill in the 1708 *Philosophical Transactions of the Royal Society*, which in 1711 he had just read because it took a few years to reach him in Hanover.

In March 1711, Leibniz sent a letter to Hans Sloane, who was the secretary of the Royal Society, complaining of the way he had been treated. The letter was read before the Royal Society on May 24, 1711, and in it, Leibniz essentially said, here we go again: "I could wish that an examination of the work did not compel me to make a complaint against your countrymen for the second time. Some time ago Nicholas Fatio de Duillier attacked me in a published paper for having attributed to myself another's discovery. I taught him to know better in the *Acta Eruditorum* of Leipzig, and you [English] yourselves disapproved of this [charge] as I learned from a letter written by the Secretary of your distinguished Society (that is, to the best of my recollection, by yourself)," Leibniz wrote to Sloane on February 21, 1711.

As he did before when Fatio had published accusations against him, Leibniz's approach was to acknowledge Newton's greatness in mathematics. Ask Newton, Leibniz essentially said—he backed me up before and he'll do so again. "Nobody knew better than Newton that this charge is false," Leibniz wrote. "For certainly I never heard of the name of the calculus of fluxions nor saw with these eyes the characters which Newton used.

"Newton himself, a truly excellent person, disapproved of this misplaced zeal of certain persons on behalf of your nation and himself, as I understand," he continued. "And yet Mr. Keill in this very volume, in the [*Transactions* for] September and October 1708, page 185, has seen fit to renew this most impertinent accusation when he writes that I have published the arithmetic of fluxions invented by Newton, after altering the name and the style of notation."

Again, as he had done in the fight with Fatio, Leibniz distinguished between Newton, whom he held in high esteem, and Keill, who was at best mistaken and at worst a liar. And, in any case, Keill had said things that needed redress. "Although I do not take Mr. Keill to be a slanderer (for I think he is to be blamed rather for hastiness of judgement than for malice)," Leibniz wrote, "yet I cannot but take that accusation which is injurious to myself as a slander. And because it is to be feared that it may be frequently repeated by imprudent or dishonest people I am driven to seek a remedy from your distinguished Royal Society."

What Leibniz wanted was for Keill to give a public statement in front of the Royal Society, retracting his accusation. Leibniz told Sloane that he wanted Keill to say that he didn't mean to say what he had said, the slander, "as though I had found out something invented by another person and claimed it as my own," Leibniz explained. "In this way he may give satisfaction for his injury to me, and show that he had no intention of uttering a slander, and a curb will be put on other persons who might at some time give voice to other similar [charges]."

On March 22, 1711, Keill appeared at a meeting of the Royal Society that was presided over by Newton, and agreed to write a letter of reply to Leibniz's demand for satisfaction. Keill prepared his response for several weeks, probably with the help of Newton, and appeared in front of the Royal Society on April 5, 1711, to present it.

Keill was unrepentant. At that second meeting, he vigorously defended himself against the libel claim in the only way possible— by prosecuting his case against Leibniz. He answered Leibniz's charges, saying that his attack on Leibniz was not without provocation but was merely a response to the anonymous review of Newton's work in 1705. He was not unfairly harsh in his criticism, he claimed, because it was a proper response to the unfair attack on Newton. Keill declared that he would produce a written account of the history of calculus and the dispute.

Keill's response was carefully crafted so that it did not accuse Leibniz of plagiarism as such but, rather, simply stated that Newton invented his calculus first, that Leibniz saw some of what Newton did, and that these "clear and obvious hints," claimed Keill, "gave him an entrance into the differential calculus."

He formally submitted this opinion in a letter to Sloane in May 1711, saying, "I have been impelled to write these lines by the publisher of the *Acta Eruditorum* of Leipzig, who in the account they have given of Newton's work on fluxions or quadrature expressly affirm that Mr. Leibniz was the discoverer of this method." Newton was the injured one, said Keill. "Whence, if I seem to have spoken pretty freely about Leibniz, I did so not with the intention of snatching anything from him but rather in order to vindicate Newton's authorship of what I take to be his own."

Finally, in a sort of courteous insult, Keill expressed amazement that Leibniz would even need to claim the invention of calculus: "Since he possesses so many unchallengeable riches of his own certainly I fail to see why he wishes to load himself with spoils stolen from others."

Keill's letter was formally presented to the Royal Society on May 24 and sent to Leibniz thereafter. Leibniz was shocked when he read Keill's response. Not only did Keill not accept his generous offer to retract his words and humiliate himself in front of the Royal Society, but he now reiterated his outrageous case even more strongly than before! This was the final straw for Leibniz. If Keill would not retract his words, Leibniz was going to shove them back down Keill's throat— or at least ask the Royal Society to make him eat them.

Although Leibniz was peeved, he did not stoop to anger. He obviosly saw himself on a whole different level intellectually from Keill and was sure that he could achieve satisfaction by getting the Royal Society (a body to which he still belonged, after all) to silence and censure Keill for his *vanæ et injustæ vociferationes* ("vain and unjust clamors"), as Leibniz perceived them.

On December 29, 1711, Leibniz wrote to Sloane again demanding redress and accusing Keill of being an upstart who was little acquainted with the facts of the case. He still had no harsh words for Newton, of course, because he respected his across-the-channel contemporary and equal. But Keill was someone for whom Leibniz had little regard—someone who was certainly not his equal.

In his letter, Leibniz said, "No fair-minded or sensible person will think it right that I, at my age and with such a full testimony of my life, should state an apologetic case for it, appearing like a suitor before a court of law, against a man who is learned indeed, but an upstart with little deep knowledge of what has gone before and without any authority from the person chiefly concerned . . . " He appealed to the society (and Newton) for redemption: "I throw myself upon your sense of justice, [to determine] whether or not such empty and unjust braying should not be suppressed, of which I believe even Newton himself would disapprove, being a distinguished person who is thoroughly acquainted with past events."

In retrospect, it seems a ludicrous approach for Leibniz to have been taking. But at the time it was entirely reasonable. During all his years of silence on the subject of calculus, Newton had never really made any aggressive public statements on the level of the ones that Keill was making. And a few years before, when Fatio had accused Leibniz of plagiarism in much the same tone, Newton had been completely silent and done nothing to defend his close friend when Leibniz had protested. Leibniz may have fully believed that Newton would back him up in his appeal to the Royal Society in his case against Keill.

Nothing could be further from the truth. In fact, Leibniz was living his last taint-free days as the widely recognized inventor of calculus. The bear trap was set, and he walked right into it. From this moment to the day he died, he would have to answer the charge that he borrowed from Newton.

What Leibniz didn't know in 1711 was that Keill *had* been discussing his accusations with Newton—that he was in fact writing with Newton's approval. In 1711 Keill had sent Newton a copy of an anonymous review of Newton's *De quadratura curvarum*, from the 1705 issue of *Acta eruditorum*, that basically implied the Englishman's original work was adapted from Leibniz's calculus—an insult that Keill was careful to point out in his accompanying letter: "I have here sent you the [article] where there is an account given of your book, I desire you will read from page 39 . . . to the end," wrote Keill.

The article was like a bucket of gasoline dumped on a campfire. Newton must have been enraged when he read the review, because he took a long time to cool down—never really cooling down until long after Leibniz died. Newton was not fooled for an instant as to the identity of the author, and he assumed from the beginning that it was Leibniz, since the *Acta Eruditorum* was the journal with which the German was so closely associated. Even though Leibniz would deny authorship of this review until the day he died, Newton guessed absolutely right: of course Leibniz had written it.

Newton drafted several responses to the review, even though he never published any of them; meanwhile, circulation of the *Acta Eruditorum* article set off a flurry of writings against him. For years, his private writings would be filled with the occasional asides and long diatribes ranting against Leibniz, who was to Newton the new Hooke, a surrogate Flamsteed, a Judas . . . Cain . . . Satan.

Newton wrote multiple drafts of a letter to Hans Sloane, commenting on the dispute between Keill and Leibniz and the now-infamous review: "I had not seen those passages before, but upon reading them I found that I have more reason to complain of the collectors of the mathematical papers in those *Acta* then Mr. Leibniz hath to complain of Mr. Keill."

Newton had a valid point. Leibniz's review was more than a little ungenerous in its assessment of his original work. But Keill's response to it, on the other hand, went for the jugular with its overt assertion that Leibniz had borrowed his ideas from Newton.

Putting on a facade of objective independence, Newton wrote to Sloane that the dispute was between Leibniz and Keill, and did not involve him: "Mr. Leibniz thinks that one of his age & reputation . . . should not enter into a dispute with Mr. Keill & I am of the same opinion, I think that it is as improper for me to enter into a dispute with the author of those papers. For the controversy is between that author & Mr. Keill." Instead of involving himself directly, Newton set the wheels of justice into motion in another way.

Leibniz would deny vehemently that he ever borrowed ideas from

Newton. His appeal to the Royal Society to decide the issue turned out to be a cataclysmic mistake, because Newton was not only the most famous and most respected scientist in this august body—he was its president. He could influence the society's disposition in the matter as perhaps no other individual could. Newton's interest was solely with Newton.

In response to Leibniz's December 29, 1711, letter and his demand for satisfaction, the Royal Society appointed a committee on March 6, 1712, to look into the matter. On paper, it was a dispute between two Royal Society members, and the society was acting in good faith and striving to fairly settle the dispute.

In actuality, there was little about the committee or its work that was truly objective. Its members were largely Newton's friends and countrymen—people like Halley. But perhaps in anticipation of the appearance of partiality toward their own countryman, several more people were appointed to the committee, including foreigners like De Moivre and Bonet, the Prussian minister.

On the strength of these appointments, Newton would later claim that that the committee was numerous in membership and international in character. Three hundred years after the fact, the claim seems flimsy, and the committee appears as if it were little more than a thinly veiled vehicle for putting forward its president's arguments. The commission did not sit down prepared to decide which was better, fluxions or calculus. They began from the premise that they were the same but for the symbols used. Hence the question of authorship became a simple matter of priority: Was Newton first?

With documents at hand (Newton's hand) proving that Englishman was first, a decision was a simple matter for the committee. What can one say about their deliberations? Their greatest achievement was that they seem to have set something of a speed record for the work of a committee.

They studied the issue for a mere six weeks, and on April 24, 1712, gave their lengthy and detailed report—a publication known as the *Commercium Epistolicum D. Johannis Collins et Aliorum de Analysi*

Promota (The correspondence of the learned John Collins and others relating to the progress of analysis.) Not surprisingly, the document found in Newton's favor and condemned Leibniz. It thrust Newton into an elevated limelight, casting him as the one who should be rightly recognized as the best mathematician in the last fifty years. It could not have been more damaging to Leibniz's reputation, painting him as a compulsive plagiarist.

"We have consulted . . . the papers of Mr. John Collins," the report began earnestly. I examined an original version of the *Commercium* at the Royal Society library in London (a reissued version from 1727). It is basically a large folder of such documents as *De analysi*, and letters to and from Collins and others, starting with Barrow to Collins in 1669 and ending with Leibniz's final 1677 letter to Oldenberg. The *Commercium* selectively abstracts pieces of these correspondences and other relevant writings with the purpose of proving that Newton was the true inventor of calculus.

The authors of the *Commercium Epistolicum* seem to have started with the premise that Leibniz was guilty, and had spent their time cobbling together fragments from letters and papers written for some forty years, to prove it. They called attention to the fact that Leibniz had a history of misrepresenting the work of others as his own—such as the affair of the eyebrow, when Leibniz had talked to the mathematician Pell and claimed as his own some of the previous discoveries of another mathematician. "He persisted in maintaining it to be his own invention by reason that he had found it himself," the committee wrote.

They also established that Newton invented calculus before 1669—as evidenced by the fact that a copy of *De Anaylsi* was found among Collins's papers.

The *Commercium Epistolicum* concluded that Leibniz had been privy to certain writings of Newton's while he was in London in 1673 and 1676, that he had received letters from Newton, and that there was no evidence that he had invented calculus before receiving those letters. It further found that Leibniz's calculus was the same as

Newton's, but for its notation, created later than the British mathematician's method of fluxions. Their decision: Keill was not libelous
and therefore need not apologize.

"We believe that those who have reported Mr. Leibniz the first
inventor knew little or nothing of his correspondence with Mr.
Collins and Mr. Oldenburg long before," the report concluded. "For
which reasons, we reckon Mr. Newton the first inventor and are of
the opinion that Mr. Keill, in asserting the same, has been noways
injurious to Mr. Leibniz."

The Royal Society and its president, Newton, accepted the report
as correct and fair, and decided to pay for its publication. While an
officially bound edition did not go on sale in bookstores, copies
became available on January 8, 1713, and the Royal Society paid for
some to be sent to key mathematicians in Europe. Several copies of
the *Commercium Epistolicum* went to Paris, and one made its way into
the hands of Abbé Bignon, who gave it to Nikolaus Bernoulli, who
carried it to Basel and showed it to his uncle Johann, who wrote
about it to Leibniz in a letter dated June 7, 1713.

The report was a stunning success, from Newton's point of view.
To him, the case was now drawn up and easily understood. It established his priority in the invention of calculus some forty years after
the fact, and did it so convincingly that, from the time the committee published its report all the way up to today, very few have mentioned calculus and Leibniz in the same breath without first
mentioning Newton.

From Leibniz's perspective, the report was a slap in the face with
a bag full of marbles. Even if one were to accept that the members
of the committee were completely objective, their conclusions are
still worth questioning. But Leibniz never had a chance to question
these conclusions because the committee extended no invitation to
the German to present his own case.

The *Commercium Epistolicum*, as flawed a document as it is,
had a profound effect on the calculus debate. It effectively subju-
gated Leibniz to a lesser status of second inventor at best, and as

opportunistic plagiarist at worst in the eyes of many. It turned the tide of popular opinion against him, and if it failed to knock him out completely, it at least knocked him onto his heels. He would spend the rest of his life fighting back but was never fully able to beat down Newton's accusations.

Leibniz's friends urged him to reply. "Most people may deduce from silence that the English case is a good one," one of them wrote. The problem for Leibniz was that Newton had cast the argument in historical terms—specifically the version of history that had him inventing fluxions long before Leibniz invented calculus. This happened to be true, and ample proof was supplied in the *Commercium Epistolicum*. But Keill had asserted that Leibniz had been given access to Newton's unpublished work, and that it had been sufficiently intelligible for him to be able to copy it. Because the *Commercium* did not attempt to disprove Keill's accusation, Leibniz was left having to prove his own innocence. In absence of a credible counterproof, the case for Newton was made all the more strong.

These were the last years of Leibniz's life, and they should have been filled with the joy of seeing his accomplishments blossoming into maturity, not a fight to retain his honor over work long past. As he never married and never had any children to surround him with grandchildren, he had to take pride in offspring of a different sort— his intellectual creations and intelligent European protégés who were inspired by those ideas to develop them further. Now Newton had taken custody of calculus, one of Leibniz's most brilliant offspring.

Leibniz was honored in 1711 with an invitation to a conference with Czar Peter the Great, who had come to Germany to see his son married to the princess Wolfenbüttel. Leibniz advised the czar at one point to open libraries and observatories in Russia, and to appoint teachers in the arts and sciences. Despite the furor in London, Peter met with Leibniz again in 1712, and he asked the German's advice on establishing and promoting math and science in Russia. Without ever setting foot in the country, Leibniz was given the title of privy counsellor of justice and awarded a nice salary. A year later the czar visited Hanover; though Leibniz was not there, he heard that Peter had had nice things to say about him.

At this time, Leibniz was not a well man. He was sick, old, and partly crippled from gout so severe that he had suffered an open lesion on his leg for two years. He ignored his leg. His attention was on matters less close to home.

11

The Flaws of Motion

▪ 1713–1716 ▪

To examine the last years of the calculus dispute does not increase one's admiration for some of the greatest of mankind.

—A. R. Hall, *Philosophers at War*

*I*n 1712, Leibniz set out to Vienna, a city that appealed to him much more than the lonely haunts of Hanover, about which he complained to his friend Thomas Burnet, "The narrow limitations, both physical and mental, within which I am confined, are owing to the circumstance that I do not live in a big city, like Paris or London, abounding with learned men from whom one can learn something, and derive some assistance." It was the last extended trip he would take in his life. He was to stay in Vienna for two years.

The Austrian capital had a lot to offer Leibniz, and it was there that he composed the *Monadologie*, one of the best-known sketches of his philosophy. He also came up with another plan for a scientific society, with all the various accompanying ambitions he was given to attaching to such schemes—a laboratory, a library, an observatory, a botanical garden, a geological collection, and a medical school. He wrote letters and memoranda in support of his vision, and wrote directly to the nobles whose support he needed most. The plan was seriously considered by the court of Charles VI in Vienna, where Leibniz had strong supporters, but they would not advance the money to carry it out.

Despite this disappointment, Leibniz was happy in Vienna. He stonewalled repeated requests in 1713 to return to Hanover after he had been in Austria for many months. The exasperated duke had his salary was frozen until he returned, and even then he delayed. The money didn't really mean much to Leibniz at that point. He had additional sources of income at this late stage in life, and was a relatively wealthy man. Upon his death in 1716, he had amassed a nest egg of around 12,000 taler, which was quite a fortune considering that an average week's wages for a common worker was about one taler.

In 1714, he was still waiting things out in Vienna, and that summer received a letter from Hanover asking whether he intended to return to the city at all. Leibniz wrote back defending himself by rolling out his record of service through four decades of the court. He might well have stayed for the remainder of his days in Vienna, however, had not fate intervened.

On June 8, 1714, Sophia was walking through the gardens of one of her homes when suddenly she was struck ill, collapsed, and died at the age of eighty-four. A few weeks after that, Queen Anne died, and suddenly it was inevitable that Sophia's son, George Ludwig, would become king of England. He left for London on September 3.

Though he took his time to make his way to it, George showed no hesitation at accepting his throne. And why should he? He was trading up from being the ruler of a small state with its seat in a second-tier European city, to becoming the monarch of one of Europe's great powers—with the fringe benefit of a new residence in one of the biggest booming metropolises in the world. At the same time, he was coming to a Britain that was rife with political infighting, social problems, and when the great South Sea bubble burst in a few years, economic ones as well.

In England, riots were commonplace, and the highways thick with robbers and cutthroats. City gates were still sometimes adorned with the rotting heads of rogues set upon pikes, and public executions—sometimes brutal stoning affairs—were considered a source of entertainment.

Everything could be found on the streets of London. Livestock—cows, sheep, chickens, and all their concomitant noises and smells—dogs barking and leaving messes everywhere; soldiers brawling and boozing at all hours; tradesmen shouting out their wares; servants rushing about; beggars and prostitutes sneering and cursing about in the foul air; elegant men and women of fortune picking their way through the cobblestone lanes with their entourages; and waste most foul draining down open sewers in the streets.

Perhaps George was the perfect king to rule over this mess, since he has been described as as coarse and crude just like so many of his subjects. According to some of the descriptions I have read, this description is perhaps even generous. He was said to be cynical, selfish, and even pathologically cruel. He may have been thrust upon the English throne, but he was to govern his own terms.

Leibniz, having heard the news of Anne's demise and knowing what it meant, proceeded back to Hanover. Surely he couldn't miss the opportunity to go to London. Even before George became king, Leibniz had hatched a plan to spend part of his time in London—to partake in conversation with the "excellent persons in whom England is so rich," as he explained to a friend.

They missed each other by three days. George, mockingly, said of Leibniz, "He comes only when I have become king." Leibniz sought to follow George, proposing to accompany Princess Caroline, but he was not well enough to travel when she left. Instead of going to London, he went to nearby Zeitz, where apparently he was introduced to a talking dog that could recite the alphabet and bark out words like "chocolate" and "coffee." The loneliness of being left behind!

In December 1714, Leibniz received a letter from Prime Minister von Bernstorff telling him not to come to London, and, a month or so later, George Ludwig, now King George I, expressly forbade Leibniz to come, ordering him to stay in Hanover until the still-incomplete history of the family was complete. Leibniz was apparently a victim of George I's advisors, who thought he would do little more than seek to interfere with their efforts in London. Leibniz continued to work for George from Hanover, producing, for instance, an anti-Jacobite pamphlet—anonymously, of course.

Leibniz's response to all this was to petition his employer to have him made historiographer of England. George I was not impressed by this request. "He must first show me that he can write history," said the king to his daughter-in-law.

Stuck in the scientific and cultural backwater of Hanover, Leibniz was now more isolated than Newton, and he remained in Hanover until he died—sick, busy, and distracted with the never-ending history project and with the calculus wars.

He should have stayed in Vienna. There, Leibniz had found the time to produce some of his best writings. Aside from the *Monadology*, he wrote an exposition on the state of philosophy and science in China, for example.

It was also in Vienna where Leibniz first heard of the *Commercium* in a letter from Johann Bernoulli, one of his biggest supporters. Bernoulli was outraged, writing to Leibniz on May 27, 1713, "This hardly civilized way of doing things displeases me particularly; you are at once accused before a tribunal consisting, as it seems, of the participants and witnesses themselves, as if charged with plagiary, then documents against you are produced, sentence is passed; you lose the case, you are condemned." He saw the *Commercium Epistolicum* as yet another blatant attempt of the British to take credit for discoveries made by intellectuals on the European continent.

Bernoulli derided Keill as Newton's ape and also wrote that he believed some of the documents in the *Commercium* to be either fabricated or altered. Worse than that, said Bernoulli, the English were accusing Leibniz of doing exactly what Newton had done: stealing the idea for calculus. Bernoulli wrote essentially that Newton had not grasped—had not even dreamt—what he claimed to have accomplished until after he had read Leibniz's work.

"Indeed, you can find no least word or single mark of this kind even in the *Principia Philosophiæ Naturalis*, where he must have had so many occasions for using his calculus of fluxions, but almost everything is there done by lines of figures without any definite analysis in the way not used by him only but by Huygens too, indeed by Torricelli, Roberval, Fermat and Cavalieri long before," wrote Bernoulli, who was absolutely right. Newton had developed the *Principia* in the old, formalized geometric style rather than with the sort of algebraic mathematics that someone using calculus would employ.

Bernoulli had made the same accusation to Leibniz years earlier, in a 1696 letter, but at that time Newton was recovering from a severe bout of depression and Leibniz was regarded in many corners as Europe's greatest mathematician. At that time, Leibniz must have seen no need to go public with such nasty accusations. Plus, Newton had never made any claim of priority in the invention of calculus, so Leibniz was probablly satisfied to leave well enough alone. But in 1713, his good name sullied, Leibniz heeded Bernoulli's words.

Upon hearing the news of the *Commercium Epistolicum*, Leibniz became convinced that it must be filled with malicious falsehoods, and he wrote to Bernoulli asking him to look into it. Bernoulli responded, on June 7, with a letter containing several pages of his opinions on the matter.

Leibniz replied a few weeks later that, while he still hadn't see the *Commercium*, he was sure that the "idiotic arguments" contained therein were worthy of ridicule. He expressed regret to Bernoulli that, for all his years of saying kind words about Newton when his comments about his counterpart were so solicited, that was the price he paid for his kindness.

He would not be so kind again, and in fact Leibniz turned very ungenerous to Newton after the *Commercium Epistolicum* appeared.

He began to question whether Newton actually had invented his own version of calculus: "He knows fluxions, but not the calculus of fluxions which (as you rightly judge) he put together at a later stage after our own was already published." In fact, said Leibniz in his letter to Bernoulli, "For many years now the English have been so swollen with vanity, even the distinguished men among them that they have taken the opportunity of snatching German things and claiming them as their own."

Bernoulli received Leibniz's letter and dashed one off in response, saying that his friend should consider proving the inferiority of the Brits by posing more challenge problems to them that could only be solved using calculus. "If such things were proposed to the English by way of a trial, it would be in my opinion the quickest way of stopping their mouths, particularly if they should reveal their extraordinary feebleness and the inadequacy of their calculus of whose antiquity they boast so greatly."

The problem that Bernoulli had posed years earlier had been a success in the sense that the only people who submitted correct answers on time were the ones who knew calculus. As Newton had proved himself equal to the challenge then, it's hard to fathom why Bernoulli and Leibniz thought a new challenge would stump him now. Nevertheless, they did seem to, and many months after Bernoulli suggested the idea, Leibniz proposed a new challenge to prove that Newton was inferior when it came to mathematics, which he included in a letter he wrote to a Venetian nobleman, the Abbé Conti. The letter to Conti ends with Leibniz stating that the purpose of the problem was, "to test the pulse of our English analysts," though the purpose of this challenge was obvious—it was clearly intended for Newton.

The challenge was to determine the curve that should cut, at right angles, an infinity of curves expressible by the same equation. Unfortunately for Leibniz, this effort failed to reveal the inferiority of the English mathematicians because there was a problem with the way the challenge was written; it was interpreted to be asking for a specific example of such a curve rather than for a general solution for finding such a curve—the much harder question that Leibniz had intended.

The general way was more difficult, requiring a mastery of calculus. But Leibniz has used some unfortunate wording that caused a number of mathematicians in England to misinterpret the challenge. Conti wrote back to Leibniz in March that "several geometers, both in London and in Oxford have given the solution."

The challenge may have been a failure, but it was not the only line of attack that Leibniz was following. In the same letter that Johann Bernoulli had sent to Leibniz telling him about the *Commercium Epistolicum*, he wrote about a mistake he had discovered a few years earlier in the *Principia*. In fact, it had been brought to Newton's attention by Johann's nephew, Nikolaus Bernoulli, who had gone to London and there met Newton in 1712.

Newton had written to Nikolaus on October 1, 1712, thanking him:"I send you enclosed the solution of the Problem about the density of resisting Mediums, set right. I desire you to show it to your Uncle & return my thanks to him for sending me notice of the mistake." He was no doubt very happy to have the correction made prior to the printing of the second edition of the *Principia* in 1713, as the revisions that went into the making of the second edition were already extensive and required years of exhaustive work.

But the fact that Newton had made this mistake in the first place may have given Johann Bernoulli cause to wonder whether the Englishman had not fully understood calculus even as late as the 1680s, when the first edition of the *Principia* had appeared. If that were the case, then Newton could not possibly be its inventor, and Bernoulli said as much to Leibniz in 1713. Bernoulli also published his criticism of the *Principia* in the *Acta Eruditorum*, the German journal closely allied with Leibniz, in this same year. However, he was reluctant to enter the spotlight and attack Newton publicly as Keill had done Leibniz. Instead, Bernoulli published this opinion anonymously.

Nevertheless, Bernoulli's doubts about Newton's abilities inspired Leibniz to write a short meditation on Newton and the whole dispute that managed, if nothing else, to stoke the flames a little more.

The *Charta Volans* (Flying Sheet) was a short printed sheet that appeared on July 29, 1713, with no author listed, though few could not guess who the author was. Like the *Commercium Epistolicum*, the *Charta Volans* was a flawed document. Leibniz, referring to himself in the third person throughout, used this paper as a vehicle to attack and mock Newton. The heart of the *Charta Volans* was Bernoulli's mistaken argument that Newton had stolen the idea of calculus from Leibniz: "After many years there was produced by Newton something that he calls the calculus of fluxions similar to the differential calculus but with other notations and terminology." It was essentially the same argument and phrases made by Newton's camp, but in reverse.

However, Leibniz made a good case for himself as the one who was duped by the other's treachery . . . because of his own trusting nature. "Leibniz on the other hand, judging others according to his own honest nature," he wrote, "readily believed the man [Newton] when he declared that such things had come to him from his own ingenuity, and so he wrote that it appeared that Newton possessed something similar to the differential calculus."

The *Charta Volans* argued that the root reason behind the position of Newton's camp in general and Keill's attack in particular was that the English suffered from an "unnatural xenophobia" that caused them to want to steal the credit from the continent and apportion the invention of calculus wholly to Newton. This would be an argument that Leibniz and his supporters would resort to over and over. It was no surprise really, as many of the figures they had known from Britain (most notably Wallis and Collins) were known to be quite protective of British accomplishments. As Bernoulli put it, in a letter to Leibniz a few months before the latter died, "It is a characteristic of the English that they begrudge everything to other [nations] and attribute all things to themselves or to their nation . . . I doubt whether you can expect [even] this much from them, that they will acknowledge Newton to be capable of error, or at any rate to have been mistaken in any one particular."

In the *Charta Volans,* Leibniz declared that once he—still in third person)—became aware of the treacherous and unfair way he was being treated, he "considered the question more carefully, which otherwise he would not have examined because he was prejudiced in Newton's favour, and began to suspect from that very procedure [of the English] which was so remote from fair-dealing that the calculus of fluxions had been developed in imitation of the differential calculus."

To support this claim that Newton had copied Leibniz, the *Charta Volans* included the "impartial" opinion of a leading mathematician who pointed out that Newton had been second to publish, and referred to the error that Bernoulli spotted three years earlier as proof that Newton's methods had been developed in imitation of Leibniz's, after the mid-1680s. Newton's supporters would later seize upon this section of the *Charta Volans* because, in addition to referencing this "leading mathematician" (which was revealed to be Bernoulli), the document referred to a certain eminent mathematician, whom they took to mean Leibniz. So Leibniz would later be mocked for calling himself an eminent mathematician.

But that summer in 1713, when the *Charta Volans* was produced, Leibniz would get the first barbs in. Perhaps the most stinging passage is where Leibniz mocked Newton's attempt to steal the credit for calculus, which the German put down to the Englishman's greed and pride: "He was too much influenced by flatterers ignorant of the earlier course of events and by a desire for renown. Having undeservedly obtained a partial share in this, through the kindness of a stranger, he longed to have deserved the whole—a sign of a mind neither fair nor honest." Moreover, the *Charta Volans* pointed to Newton's earlier troubles with Hooke over the *Principia,* and to his falling out with the astronomer John Flamsteed over his theory of lunar motion: "Of [Newton's tendency not to give others full credit], Hooke too has complained, in relation to the hypothesis of the planets, and Flamsteed because of the use of [his] observations."

Leibniz got one of his friends, a man named Christian Wolf, to print and circulate the *Charta Volans* for him. By early 1714, copies were being spread around Europe, and Johann Bernoulli wrote to Leibniz that May to share the good news: "Mr. Wolf has sent me many copies of the sheets containing your reply (for Wolf has said it is yours, and the statement appears publicly in the German journal, *Büchersaal*, which is printed in Leipzig; and has asked me to distribute it amongst the mathematicians known to me; of course I have already done so, and I have especially sent quite a number into France; but I was reluctant to send any to England, lest the English suspect that I am author of that reply."

From there, the controversy and the battle increased in intensity. Though Leibniz denied that he was the paper's author, few (and least of all Newton) doubted where it came from. Newton was sent a copy by a man named John Chamberlayne, and, after incredulously reading the *Charta Volans,* he became almost obsessive in his pursuit of the case against Leibniz. Newton wrote a number of drafts of responses, several of which were found among his papers when he died, though he ultimately never published them nor sent them to others in the form of letters.

Meanwhile, in the summer of 1713, a new Dutch journal, *Journal Literaire de la Haye*, was launched that carried a translation of the *Commercium Epistolicum* (done by Leibniz's man Wolf) in its first issue. Playing to both sides of the fence, the journal also published a paper called the "Letter from London" that was written by Keill, which included an extract of a letter Newton had penned to Collins more than forty years before, in which Newton described his method for finding tangents. Keill claimed that the same letter had been sent to Leibniz. To this, Leibniz responded with an article, "Remarks on the Dispute," at the end of the year. In it, he again touted the evidence that the mistakes made in the *Principia* supposedly proved.

Continuing to pay special attention to the dispute later that same year, another issue of *Journal Literaire de la Haye* reprinted the *Charta Volans* plus an anonymous review of the *Commercium Epistolicum* by

Leibniz, along with an anonymous response to Keill's remarks that was also written by Leibniz.

The reason for all this anonymity was simple: For Leibniz, the fact that Keill was attacking him was not acceptable. He did not seem to want to engage in a head-to-head fight with someone who was not only much younger and much less accomplished than he was, but who was fundamentally a less intelligent mathematician. Leibniz seemed to feel no need to denigrate himself by replying directly to an underling like Keill—but rather aimed to take on Newton directly.

But Newton and Keill already has a nice working arrangement, and neither was about to disrupt it. Keill wrote to Newton on February 8, 1714, telling him of the review of the *Commercium* and asking him, "I would gladly have your opinion what you think is needful further to be done in answer to Mr. Leibniz . . . I am of opinion that Mr. Leibniz should be used a little smartly and all his Plagiary and Bluders showed at large." Then Keill wrote another two letters to Newton on the subject, saying of Leibniz's remarks that he "never saw any thing writ with so much impudence falsehood and slander," and that they must be answered immediately.

Newton replied casually, almost two months later, "If you please when you have it, to consider of what answer you think proper, I will within a post or two send you my thought upon the subject, that you may compare them with your own sentiments & then draw up such an answer as you think proper." Newton wrote no fewer than seven drafts of a reply to Leibniz's anonymous "Remarks" but never published any of them.

Instead, it was up to Keill. He sent Newton a draft of his answer in May, and it eventually grew into a forty-two-page article, which he sent to the *Journal Literaire de la Haye* for publication in their July/August 1714 issue. There is good reason to believe that Newton had played a major role in this "Answer to the Author of the Remarks," as the article was called, as it was written at a high enough level to probably have been over Keill's head.

Now that the dispute was fully out in the open and there were numerous published accounts of it, many more people were becoming aware of it, and a number of contemporaries of both men couldn't help but to get involved. Newton's enemies among the English intellectual elite, for instance, would send Leibniz copies of such publications as the *Commercium Epistolicum*, as well as word on what Newton was up to. The astronomer John Flamsteed sent Leibniz a list of errors in Newton's lunar theories.

To some of Leibniz's supporters, the *Charta Volans* was not enough. If Leibniz could respond directly to Newton with his own *Commercium Epistolicum*, his case would be greatly bolstered. Bernoulli suggests that doing so would bring about a sound victory. "I think that Mr. Newton will some time smart for so easily lending his ear to flatterers," Bernoulli wrote. "Meanwhile it will be wise for you to concentrate on your reply to the *Commercium Epistolicum*, finish it in good time and lay it before the public, lest they should have reason to rejoice in the delay."

Indeed, Leibniz made noises that his own *Commercium Epistolicum* would be more fair because it would include all the relevant letters and documents, insinuating that Newton had hand-chosen certain documents while ignoring others. When Newton heard of this criticism he said that if Leibniz had letters to produce, then he should go ahead and produce them. He added that there were even more damning letters than the ones that were included in the *Commercium Epistolicum*, and these were not published.

Leibniz wrote to Johann Bernoulli, toward the end of 1714, "Many distinguished men there [in England] do not at all approve the boldness of Newton's toadies . . . I am resolved to publish some correspondence of my own, from which it will appear how weak Newton once was in other respects."

But this was not the easiest thing for Leibniz to do. First of all, he was in Vienna from 1712 to 1714, and far out of range of access to all the relevant letters. Second, it would not have been easy for him to go through his papers and come up with only the most relevant bits—he had massive piles of correspondence stretching back over decades. Going through a stack of these letters would not have been so simple as simply flipping through documents already gathered toward an express purpose, as the committee that had assembled the *Commercium Epistolicum* had done. Moreover, many of the German's papers were a mad scramble of tiny writing, some so small as to be barely legible without a magnifying glass. Add to this marginal notes in the same hand and multiple corrections—additions, deletions, and word changes . . . even to Leibniz, familiar with his own words, the unlikeliness of a quick skim through must have felt hopeless. And there, all the time in the background, was the pressure his employer continued to exert on him to finish the historical work.

Meanwhile Newton must have recognized that the *Commercium Epistolicum* might not be enough to support his own case. He wrote a paper called "An account of the Book entitled *Commercium Epistolicum*," in 1714, and published it anonymously in the January–February issue of the *Philosophical Transactions of the Royal Society*. It filled all but three pages of the issue. He further had it translated into French and published it in the *Journal Literaire de la Haye*, arranged for a review of it to appear in another journal called the *Nouvelles Litteraires*, and had it printed as a separate pamphlet and had it distributed through Europe. Then, for good measure, he had it translated into Latin. Finally, Newton was the prolific author his contemporaries had wanted him to be for so many years.

In this "Account," Newton attacked and devalued one of Leibniz's greatest contribution to mathematics: his invention of the symbols of calculus, which had greatly enhanced the ability of mathematicians to learn and apply the methods of calculus that are still in use today. Newton, wrote Newton hautily, does not confine himself to symbols.

Feelings were equally hostile in the Leibniz camp, and Leibniz's supporters generated a great deal of ill feeling—much of it directed at Keill. Christian Wolf wrote a letter to Leibniz in the second half of 1714 that complained of the man and his childish reasoning: "I wonder at the impudence of the man, and also I wonder at his boasting . . . that he fight not with his own weapons, but with Newton's."

Leibniz replied to Wolf several months later: "I cannot bring myself to make a reply to that crude man Keill. I have held what he has put forward hardly worth reading." In another letter, Leibniz showed even more of his true feelings: "Since Keill writes like a bumpkin, I wish to have no dealings with a man of that sort. It is pointless to write for those who respond only to his bold assertations and boasting, for they do not examine the substance . . . I think of knocking the man down, some time, with things rather than words." While Keill was several years younger and not crippled by gout as Leibniz was, my money would have been on Leibniz—angry as he was.

Leibniz, at this point, was desperate to bring Bernoulli into the fray so that he could champion him the way that Keill was championing Newton. Bernoulli was the perfect man to play that role. He was a master of calculus and had been using it for decades. He was also a very distinguished mathematician, unlike Keill who was secondary in skills and accomplishments to Newton and Leibniz both. In fact, Bernoulli was one of the few people alive who was the mathematical equal to both parties in the dispute—and perhaps even more brilliant and pure a mathematician than either man.

Bernoulli would have made a much more formidable second than Keill, and his disposition was perfect for Leibniz. He firmly came down on Leibniz's side in the matter, and he was already a "leading mathematician" whose anonymous criticism was contained in the *Charta Volans*. So why not bring him out in the open?

Bernoulli did not want to be on the front lines of the calculus wars, and he asked Leibniz to keep him out of the controversies. Bernoulli did not want to have his name associated with the dispute because he was torn. On the one hand, he was loyal to his friend and longtime collaborator—Bernoulli's own career as a mathematician was advanced as a result of his picking up the threads that Leibniz had spun and weaving calculus into a set of mathematical tools that could be grasped and applied by many mathematicians. At the same time, Bernoulli wanted to be diplomatic in his direct dealings with

Newton because he personally harbored no ill will toward England's greatest scientist. In fact, he must have felt the opposite—Newton was the friendly colleague who had helped Bernoulli gain admission into the Royal Society, and also had been the gracious host who had entertained Bernoulli's son when he was in London.

Still, Leibniz was not going to accept no for an answer that easily. He did little to conceal Bernoulli's true allegiance, and once, writing a letter referring to the most recent mathematical challenge that had been proposed "to test the pulse of the English analysts," he outed Bernoulli as the one who had conceived the problem. He also sought to draw Bernoulli out by telling him that Newton knew the letter referred to in the *Charta Volans* was his. "I wonder how Newton could know that I was the author of the letter," Bernoulli wrote back, "since no mortal knew that I wrote it except [you and I]."

Finally, Leibniz let slip that Bernoulli was the author of the letter referred to in the *Charta Volans*, when he anonymously reviewed Newton's "Account of the *Commercium Epistolicum*" in 1715. To draw Bernoulli out, Leibniz also began naming him in correspondences as one of Newton's critics.

Once Newton found out that Bernoulli was the mysterious "eminent mathematician," he wasted no time in insulting him, calling Bernoulli a "pretended" mathematician in 1716.

Bernoulli would deny his authorship of this letter for years, and after Leibniz died, sought to make amends, letting Newton know that he was not the author and that Leibniz had been misled in attributing it to him. He wrote to the French mathematician Pierre Rémond de Monmort, "I desire nothing so much as to live in good fellowship with him, and to find an opportunity of showing him how much I value his rare merits, indeed I never speak of him save with much praise."

Newton, for his part, accepted the olive branch from Bernoulli, writing to Monmort in France, "I readily welcome and court his friendship."

THOUGH BERNOULLI BALKED at getting between Newton and Leibniz, there were many others who were more than willing to do so—and not solely because they were advocating for one man or the other. Indeed, as tempers flared and hostilities became more and more open, many third parties on both sides of the English Channel were anxious to see the dispute reach an amicable conclusion.

The ambitious John Chamberlayne, who was in correspondence with both Newton and Leibniz, tried to single-handedly settle the dispute. He sent a letter to Leibniz, then in Vienna, on February 27, 1714, telling him, "I have been inform'd of the differences fatal to learning between two of the greatest philosophers & mathematicians of Europe, and I need not say I mean Sr. Isaac Newton and Mr. Leibniz, one of the glory of Germany the other of Great Britain, and both of them men that honor me with the friendship which I shall always cultivate to the best of my power, tho' I can never deserve it … yet as it would be very glorious to me, as well as advantagious to the commonwealth of learning, if i could bring such an affair to a happy end."

But Chamberlayne's desire to make harmonious wine of the vinegary dispute would die on the vine. Really, all that his efforts did for Leibniz was to allow him yet another outlet via which to vent his anger. Leibniz wrote back, in April 1714, in a harshly-worded letter, that Newton's purpose in bringing out the *Commercium Epistolicum* had been to unfairly discredit him, and that he doubted whether Newton had invented calculus at all before reading Leibniz's work. Nor was Newton any more willing to let bygones be bygones. Chamberlayne sent news of Leibniz's letter to him, and Newton replied that he would not retract things that were true and that, because the *Commercium Epistolicum* was a true document, it in no way did Leibniz an injustice.

Leibniz wrote another letter to Chamberlayne in which he laid out his dissatisfaction with the *Commercium*, asking the Englishman to submit this letter to the Royal Society. It reads in part: "I do not at all believe that the judgment which is given can be taken for a final judgment of the Society. Yet Mr. Newton has caused it to be published to the world by a book printed expressly for discrediting me, and sent it into Germany, into France, and into Italy as in the name of the society. This pretended judgment, and this affront done without cause to one of the most ancient members of the Society itself and who has done it no dishonor will find but few approvers in the world."

Newton translated this letter himself and had it read before the Royal Society. The members snubbed this effort, however, passing a resolution to coldly ignore the letter without comment. The journal of the royal Society records on May 20, 1714, "The Translation of [Mr. Leibniz's] Letter to Mr. Chamberlayne produced the last meeting was read. It was not judged proper [since this letter was not directed to them] for the Society to concern themselves therewith, nor were they desired to do so . . . "

Keill, on the other hand was more than willing to take on whatever Leibniz had to offer, writing to Chamberlayne a few months later, "If Mr. Leibniz makes any more noise I will still give the world a greater knowledge of his merits and candor."

This created such animosity among Leibniz, his supporters, and Keill that they began to regard the latter in the cruelest manner. For example, Leibniz's friend Wolf wrote him a letter in which he spread the most stinging gossip about Keill: "A few days ago I learnt from someone from England who visited me that Keill had behaved so unlike the occupant of a professiorial chair because of his disgraceful morals (for he has frequented drinkshops and bawdy-houses with the students entrusted to his care, spending heavily on wine and women) that he may become notorious for some infamous proceedings arising from his want of morals. . . . "

Even while such whispers were spread against Keill, further evidence was spread against Newton. Leibniz wondered aloud in his letters, some of which he expected Newton to be privy to, about a famous paragraph in the *Principia* that was in the first edition but that Newton had retracted from the second.

He also wrote to such people as the Abbé Conti and a Madame de Kilmansegg, saying that Newton had accorded to him the invention of calculus years earlier, in the second lemma's ending scholium of the second book of the *Principia*. In this paragraph, Newton wrote: "In a correspondence which took place about ten years ago between that very skillful geometrician, G. W. Leibniz, and myself, I announced to him that I possessed a method of determining maxima and minima, of drawing tangents, and of performing similar operations, which was equally applicable to rational and irrational quantities, and concealed the same in transposed letters . . . This illustrious man replied that he also had fallen on a method of the same kind, and he communicated to me his method, which scarcely differed from mine except in the notation."

Strangely, this scholium, as the passage is called, had different meanings to Leibniz and his supporters than it did for Newton and his. Leibniz seemed to take this to mean that Newton was admitting that Leibniz was in possession of a method like Newton's own. Newton and his supporters looked at it as establishing his priority as the inventor.

This difference of opinion was reflected in the pages of another book, *History of Fluxions*, by British mathematician Joseph Raphson, which appeared in 1715 to further Newton's cause.

Raphson, who had died before his book hit the streets, had reviewed a half dozen previous published documents that were available to him. Although Newton's work was not yet published or available to the public, he had allowed Raphson to read some of his personal papers periodically, through the years. The book was clearly biased toward Newton, reiterating in its preface that Newton had the priority and the genius both. Raphson went even further than seeking to set the record straight, by establishing a chronology in favor of Newton, suggesting at the same time, perhaps unfairly, that Leibniz's calculus was "less apt and more laborious" than Newton's.

Newton wrote a densely typeset seven-page supplement to the

book in which he defended his earlier words in the scholium and claimed that it was a matter of misinterpretation on Leibniz's part rather than any admission on his part: "It was written not to give away that lemma to Mr. Leibniz but, on the contrary, to assert it to myself."

When Leibniz became aware of the *History of Fluxions*, he was already writing his own version of the history, calling it the *History and Origin of the Differential Calculus*. This was no new idea. Twenty years earlier, he had written to Huygens with essentially the same intention, to write a book on calculus (albeit one that was a little more forward looking). "Your exhortation confirms me in the purpose I have of producing a treatise explaining the foundations and applications of the calculus of sums and differences and some related matters," Leibniz wrote. "As an appendix I shall add the beautiful insights and discoveries of certain geometricians who have made good use of my method, if they will be so kind as to send them to me. I hope that the Marquis de l'Hospital will do me this favor if you judge it fitting to suggest it to him. The Bernoulli brothers could also do it. If I find something in the works of Mr. Newton which Mr. Wallis has inserted in his algebra which will help get us forward, I shall make use of it and give him credit."

But, as the Brunswick history, Leibniz never finished it. He may have lacked the patience that was needed to carefully comb through his old notes and letters or he may have simply been too busy with his other things. Nevertheless, His fragmented *History and Origin of the Differential Calculus* is a document that's both beautiful and jarring. The opening paragraph, which I quoted at the beginning of chapter 5 of this book, is an outstanding statement of the importance of recording a discovery of any sort—particularly one of the importance of calculus. "Among the most renowned discoveries of the times must be considered that of a new kind of mathematical analysis, known by the name of the differential calculus; and of this, even if the essentials are at the present time considered to be sufficiently demonstrated, nevertheless the origin and the method of the discovery are not yet known to the world at large . . . ," Leibniz wrote.

Then, in the paragraphs that followed, the *History* became much more bitter and mired in the dispute at hand:

> Now there never existed any uncertainty as to the name of the true inventor, until recently, in 1712, certain upstarts, either in ignorance of the literature of the times gone by, or through envy, or with some slight hope of gaining notoriety by the discussion, or lastly from obsequious flattery, have set up a rival to him; and by their praise of this rival, the author has suffered no small disparagement in the matter, for the former has been credited with having known far more than is to be found in the subject under discussion. Moreover, in this they acted with considerable shrewdness, in that they put off starting the dispute until those who knew the circumstances, Huygens, Wallis, Tschirnhaus, and others, on whose testimony the could have been refused, were all dead.

Leibniz was in the middle of this *History* when he received a letter from Newton himself. This letter was the fruit of another effort to broker peace—ultimately not successful except that it led to one final exchange of letters between the two. It started when Newton had Abbé Conti arrange for the ambassadors and foreign ministers who were in London, including Baron de Kilmansegg, the ambassador from Hanover, to assemble and decide the issue for themselves. It was a confident and bold move, but one that was doomed for failure. While the ambassadors were more than happy to gather to discuss the dispute, they were not able to come to a decision.

I'm not surprised, really. Newton had arranged for them to view the *Commercium Epistolicum* and related papers for themselves. But these were no easy documents for anyone to peruse, let alone an international group of nonmathematicians who would have nevertheless prided themselves on their intellectual abilities, or at least interests, such that their need to save face would have prevented their admitting they were not up to the task.

As a solution, the baron urged the English mathematician to write to Leibniz himself, which the Abbé Conti reported back to Newton. Since he had been the one to arrange for the ambassadors to decide the issue, Newton had to follow through with a letter. He did so on February 26, 1716, which the Abbé Conti forwarded to Hanover.

Newton apparently spent many hours drafting this letter, though there was nothing new in it. It was yet another bitter rehash of all the evidence. To him, the *Commercium Epistolicum* was a factual and fairly collected pile of evidence published "by a numerous committee of gentlemen of several nations." He displayed no intention of retracting a word of it.

Newton probably had the sense that his argument, solid thus far, was worth sticking to. The letter contained some criticisms of Leibniz's philosophy, and then ended by stating that it was up to the German to prove his accusations of plagiarism against Newton. "But as he has lately attacked me with an accusation which amounts to plagiary; if he goes on to accuse me, it lies upon him by the laws of all nations to prove his accusation . . . he is the aggressor & it lies upon him to prove his charge," Newton wrote.

The Abbé Conti wrapped Newton's letter with his own; in this cover letter, he asked Leibniz directly who invented calculus first. Leibniz wrote to Bernoulli soon after, gloating, "Newton himself, since he saw that I regarded Keill as unworthy of an answer, has entered the ring, having written a letter to the Abbé Conti, who has sent [it] to me."

Bernoulli replied to Leibniz, "It is a good thing that Newton has at last entered the ring himself, in order to fight under his own name, and laid aside his mask . . . Whatever it may be, I hope now the historical truth will be more clearly discovered, if only Newton will, with that candor which I suppose and trust him to possess, tell faithfully the things which have happened, and will publicly acknowledge the truth of what you have put forward."

But the exchange would not bear such hopeful fruit. Newton, after getting Leibniz's letter, responded with an even longer letter containing more reiteration.

Then Leibniz, perhaps sensing that he was finally beginning to confront Newton head-to-head as he had long sought, did the eighteenth-century equivalent of posting his opinion on a public Web site. Seeking to bring as many people into the fray as he could, he sent copies of the correspondence to Paris to be shared and distributed. Leibniz sent his response through Rémond de Montmort—telling him that it was a letter that he wanted communicated to all the mathematicians in Paris in order that they could all be his witnesses. In the letter-proper, Leibniz denied the accusation that he was the aggressor who was accusing Newton of plagiarism, and again blamed the influence of those who would flatter him. "The wicked chicanery of his new friends has greatly embarrassed him," Leibniz wrote of Newton.

Newton's "Observations" on Leibniz's letter, which he recorded shortly thereafter, show how bitter he had become. "Mr. *Leibniz* accuses them [the committee appointed by the Royal Society] for not printing the letters entire (including as well what did not relate to the matter referred to them, as what did relate to it,) as if it were not lawful to cite a paragraph out of a book, without citing the whole book. Thus he complains, that the *Commercium Epistolicum* should have been much bigger. But when he is to answer it, he complains that it is too big, and would require an answer as big as itself."

Where this might have led is anybody's guess. But the correspondence did not continue. Instead, Leibniz stepped back from any discussion of calculus to attack Newton's worldview—that is, Newton's understanding of gravity. Here, Leibniz was no doubt sure, his rival was weak, because the Englishman believed in the hard-to-grasp and impossible-to-justify notion of universal gravitation—action at a distance. Like many of his other contemporaries, Leibniz had difficulty accepting Newton's theory.

Leibniz had prefaced this attack in a letter to Bernoulli. "Newton in no way demonstrates by means of his experiments that matter is everywhere heavy, or that any part whatever is attracted by any other part, or that a vacuum exists, in accordance with his own boasts," he had said.

Leibniz clearly wanted to shift the entire debate onto more philo-sophical grounds. He was, after all, one of the most preeminent philosophers in Europe (a distinction Newton could not claim) and he perceived his advantage in this regard. "His philosophy seems rather strange to me," Leibniz wrote of Newton to the Abbé Conti. "I do not think it can be established."

This was not something that Leibniz did whimsically. He prob-ably really thought Newton was wrong, and he must have been convinced that what he saw as Newton's ill-founded natural phi-losophy would sink him.

NEWTON WAS CONVINCED that there existed what we would today call a Newtonian universe—that gravity obeying deterministic laws governs all matter. In his earlier days, he had worked out his theory of universal gravitation as a way of describing things like the tides and the motion of the planets around the sun. He didn't attempt to explain what gravity was, but rather satisfied himself and his readers by describing how it worked.

Gravity, for Newton, can best be understood by the equation he created to describe it. The force due to gravity that is exerted by two objects on each other is a function of the masses of the two objects and the inverse square of distance between them. It was, for New-ton, a force that stretched across empty space.

Across the English Channel, Leibniz had profound problems with Newton's physics because he was at heart a rationalist. He was per-fectly willing to accept the mathematical formulation that gravity was inversely proportional to the square of the distance between two objects, but this purely mathematical formulation of reality was not enough for Leibniz. He needed it to be rational.

To Leibniz, one of the principles upon which science was founded was that of sufficient reason: that nothing happens without a reason suf-ficient for it to happen. He once wrote, "The fundamental principle of reasoning is, nothing without cause." He also wrote, "This axiom, however, *that there is nothing without a reason*, must be considered one of the greatest and most fruitful of all human knowledge, for upon it is built a great part of metaphysics, physics, and moral science."

Leibniz probably did not like Newton's theory of universal gravitation on the simple premise that action at a distance (as in gravity, exerting a force even through the separation of millions of miles) had to be impossible. He outright rejected the theory as absurd. Or as Leibniz expressed it coldly, "I believe that one must have recourse to a kind of perpetual miracle to explain this effect."

The once prevailing theory to which universal gravitation was an alternative was the notion that the planets are carried around the sun in vortexes, and Leibniz was a firm subscriber to this theory because it made much more sense than some mysterious miracle force called . . . what was it? . . . gravity?!

For him, the reason for the motion of the planets was simply one of matter—that is, the matter surrounding the planets pushing on the matter that is the planets. Leibniz looked at the fact that all the planets are in the same plane as the sun and reasoned that it was because they were spinning around in a massive vortex of matter. This motion is like the movement of a leaf in a stream, carried along by the billions of water molecules, and, just as without the water the leaf could not float downstream, without the vortex matter "nothing would prevent the planets from going in every direction," he wrote.

This theory was robust and was employed by Leibniz to explain other things, such as the round shape of the earth, in a very convincing imagining: "If a body is surrounded by another which is more fluid and more agitated, to which it does not permit a sufficiently free passage into its interior, it will be struck from without by an infinity of waves which will help to harden and to press its parts together. A spherical body is less exposed to the blows of this surrounding fluid, because its surface is the smallest possible and because the uniform diversity of its internal motion as well as the external motions contributes to this roundness."

Newton was likely, of course, enraged by what he regarded as Leibniz's attempt to change the subject. He probably had no desire to enter into an extensive argument with Leibniz over natural philosophy, and he was spared from having to do so. Instead, another one of Newton's proxies took up the debate.

Leibniz wrote letters that were critical of Newton's worldview to Caroline, the Princess of Wales, in November 1715. She was the daughter-in-law of George Ludwig, who by then was sitting on the throne as England's George I, and was someone who was familiar to Leibniz and somewhat of a champion of his philosophy and person. She passed the letters on to a man named Samuel Clarke, who was uniquely positioned to argue Newton's worldview with Leibniz. Clarke, the king's chaplain, had translated the book *Opticks* into Latin in 1706 for a large fee, and a decade later, was asked by Princess Caroline to translate Leibniz's *Theodicy* into English. This he refused to do, but he did respond to Leibniz in writing.

In a letter to Caroline, Leibniz criticized Newton for relying upon divine intervention to explain phenomena and to maintain the function of the universe. Newton's universe, as he saw it, was a badly constructed clock in need of occasional repair. He objected to this sort of need because he professed belief in the uniform rationality and morality of the universe, and he expressed that God's decisions were behind everything. Those decisions, he believed, were derived from the same principles of rational and moral human decisions. Clarke responded by arguing against Leibniz, and this began one of the most famous exchanges in the history of philosophy—the so-called Leibniz-Clarke correspondence. This exchange, though short-lived, was significant enough to be published almost immediately, in 1717, and continues to be published today.

Ultimately, though, Leibniz's attempt to draw Newton into an argument on either metaphysical or philosophical grounds amounted probably to less than he had hoped. Newton never took the bait and there was never a direct discussion between them on the subject of matter. Furthermore, while this may have been a smart and obvious choice for Leibniz to attempt at the time, it was a poor decision historically because his attack on Newton's theory of universal gravitation weakened his own argument.

Despite the fact that Leibniz clearly saw himself to be on much higher ground, Newton was right about gravity. The arguments Leibniz made against him are somewhat embarrassing historically, because it was one area in which this brilliant man was dead wrong. As the eighteenth century wore on and after both men died, the balance of opinion was to sway in favor of Newton, and the scientists and mathematicians who followed these men began to realize more and more the reality of gravity. The theory of vortexs, while it had its supporters even into the eighteenth century, was destined for the dustbins of science.

And as gravity emerged triumphant, so too did many writers emerge to champion Newton. Perhaps the most famous of these was Voltaire, who derided the theory of the vortex and celebrated Newton for his of gravity. "Sir *Isaac Newton* seems to have destroy'd all these great and little vortices," he wrote. And he added, "This power of gravitation acts proportionally to the quantity of matter in bodies, a truth which Sir *Isaac* has demonstrated by experiments."

A consensus was reached years after Leibniz and Newton died: Because Newton was right on gravity, perhaps too, many must have thought, he was right on the true origins of calculus as well. Thus, it was an unfortunate side skirmish in the calculus wars that Leibniz chose to support his case by attacking Newton on gravity.

Purged of Ambiguity

Death troubles himself neither with the execution of our projects, nor with the improvement of science.

—Leibniz, from a letter to Thomas Burnet, 1696

Toward the end of Leibniz's life, as the battle with Newton was reaching full throttle, it had the potential to take on an increasingly political tone, as his boss was now King of England. But anyone who might have assumed that George I would have more reason to side with Leibniz was completely wrong. Newton was a Whig, and the Whigs were generally loyal to the House of Hanover, so Newton was surely okay as far as George I was concerned.

In fact, the attitude of George I toward the calculus dispute seemed to be one of indifference—not so much out of a lack of interest but more an indifference that comes from knowing that, regardless of whom was right in the dispute, he was lord of both participants. "I think myself happy in possessing two kingdoms, one in which I have the honor of reckoning a Leibniz, and in the other a Newton, among my subjects," he once said.

Besides, George had a strange relationship with Leibniz, ever leaning on him to stop stalling and complete the history of his family. Because of this and other reasons, Leibniz spent his dying days in Hanover while George and most of his court were in England—an abandonment perhaps, or something that shows lack of favor at the very least. Perhaps more revealing of their relationship is an incident that occurred in 1711. When Leibniz injured himself in a fall that year, sick, old, and partly crippled man that he was, George is said to have been amused and even saw it as fitting that it happened. He fell well short of benevolence toward his family's longtime employee.

The injury was just one in a long line of physical insults that Leibniz would endure in the final years of his life. Leibniz was suffering from gout, which is an extremely painful form of arthritis caused by the buildup of needlelike crystals of uric acid in the connective tissues and joints. These buildups cause inflammation and shooting pains in the joints, and such attacks can take days to subside. Toward the close of Leibniz's life, his gout worsened. "I suffer from time to time in my feet; occasionally the disease passes into my hands; but head and stomach, thank god, still do their duty," Leibniz wrote in 1715.

He also developed a nasty abscess in his right leg that made it difficult for him to walk, perhaps because of his tendency toward a lack of movement. He is said to have often been given to sitting for hours—sometimes days on end—working from his chair.

Nevertheless, he never let the pain get the best of him. He would deal with the attacks by lying perfectly still in bed and at times by tightening wooden vices around the affected joints. Unfortunately this, apparently, damaged his nerves so badly that he became permanently bedridden.

In November 1716 he lay in bed for eight days, finally agreeing to see a doctor, a Dr. Seip, on Friday the thirteenth. One history paints an interesting picture of the patient as a living encyclopedia, with an in-depth knowledge of the art and application of medicine, discussing alchemy and history with the doctor while he was wracked with pain and his pulse weakened. Leibniz had broken into a cold sweat across his forehead and was perspiring profusely. He was shaking uncontrollably, surrounded by books and notes and other work, and though he tried to work, he could not write anything.

The doctor gave a dire prognosis: Leibniz would surely have no chance of recovery. He gave him some medicine. Leibniz lasted through to the next day, and on November 14, 1716, this most famous son of Leipzig died in his reluctant longtime residence of Hanover.

His coffin had to be built, and Leibniz's secretary, Eckhart, ordered an ornately designed and expensive one that was decorated with lines from Horace, symbols of mathematics and rebirth. The funeral was a couple of days later, after which Leibniz was transferred to the Neustädter church where he was to be buried. He was buried *inside* the church, which was rare for a commoner back then. There is a sandstone marker with the inscription "Ossa Leibnitii" over what are today believed to be his remains.

Leibniz's star grew in brightness after his death. In the eighteenth century, he was regarded as a very important intellectual, and a monument was erected in his honor around 1780, which was again extremely rare for a non-noble. This is described as a circular temple with a white marble bust in the middle and the inscription "Genio Leibnitii." A measure of his worth was that, years later, when the church was renovated, the bones of the people buried inside it were exhumed. Only Leibniz was reburied within the renovated structure.

Still, many historians have commented on the paltry attendance at his funeral. A man named John Ker, of Kersland, who happened to arrive in town the day that Leibniz died, was apparently struck by the lack of attention paid by the locals. He commented, apparently, that Leibniz was buried more like a common thief than one of the ornaments of his country.

Most of George's court was in London but the king and his entourage were hunting nearby when word reached them of Leibniz's death. History records that, despite the fact that the entire court had been invited, the members of the court, most notably George I himself, did not attend it.

Several obituaries appeared in honor of Leibniz. The *Journal des Savants* published an account of his death in 1717, and another publication in the Hague appeared in 1718 with an "Éloge historique de M. de Leibniz." The Académie des Sciences in Paris took notice, and the secretary there wrote an eulogy to Leibniz that he read to the members in 1717.

The Royal Society gave no notice of Leibniz's passing, however, even though he was still a member. But perhaps the greater insult was that the Society of Sciences in Berlin did nothing to mark the occasion, even though Leibniz had been its first president and founder.

Shortly after Leibniz died, the Abbé Conti wrote to Newton to inform him of the fact. "Mr. Leibniz is dead," Conti wrote, "and the dispute is finished." But it was not nearly over for Newton.

As soon as Newton heard that Leibniz was dead, he pushed a reissued edition of Raphson's book into print, and into this he inserted his own words in response to the letter Leibniz had sent him. Newton's feelings toward Leibniz did not seem to soften with the passing years, not even after the death of his archrival. Two years later, the Englishman wrote a long, gloating passage about how Leibniz had never been able to refute his arguments. He continued to write bitter letters and treatises for years after Leibniz's death, though he kept many private and those were not discovered until after his own death a decade later.

The letters that were in his possession when he died reveal how deeply wronged he felt by the whole affair, that he had been unfairly treated by Leibniz. He maintained to the grave that Leibniz was the aggressor and he, Newton, was the one who was defending himself from accusations of plagiarism. There can be but one true inventor of anything, Newton insisted, regardless of who improves upon the invention.

He was quite successful in spreading his belief in his greatness to the detriment of Leibniz's—as were his followers. Voltaire, of course, was Newton's greatest champion in France. After spending a few years in England, he wrote a number of essays that extolled Newton and Newtonianism, including one of the first popularizations of the Englishman's ideas. Voltaire was rather unsparing in his treatment of Leibniz and his philosophy, many years after the older man had died. Leibniz was spoofed and ridiculed by Voltaire as the silly Dr. Pangloss of the novel *Candide*. His very name, Pangloss (broad summary), is a reference to the philosophy that came to oversimplify Leibniz's outlook after he died—the notion of the best of all possible worlds.

Leibniz theorized that the total exclusion of evil in the world was impossible but that humans did live in the best of all possible worlds in the sense that the least amount of evil was allowed. Leibniz was not saying by "the best of all possible worlds" that every aspect of the world was perfectly without flaws. He was a witness to too many wars and too much suffering to think anything that stupid. All he was really saying was that, of the infinite number of possible worlds, this was the best. Suffering and the horrors of the world were part of a larger order, in Leibniz's view, that remained harmonious. Moreover, he argued that the universe must be imperfect, because otherwise it could not be distinct from a perfect creator.

Though Leibniz was ridiculed by Voltaire's cursory mocking of his philosophy, Bertand Russell, who wrote one of the definitive expositions on Leibniz's outlook, called it an unusually complete and coherent system. But, however admired by Russell Leibniz would become, and however simple and elegant a concept his best of all possible worlds was, its Hollywood-style simplicity came to represent Leibniz's philosophy after he died, and the phrase "the best of all possible worlds" became a mantra that was to tar and feather many aspects of Leibniz's work in the eighteenth century and beyond. In the years immediately following and for centuries, he suffered from the perception that he was overly optimistic—that he was, in the words of one historian, the best of all possible worlds.

Even in the twentieth century, the best of all possible worlds is still the subject of some amusement. In Woody Allen's *Love and Death*, Diane Keaton's character holds up two dried, perfect leaves, comments on their beauty, and says that their beauty demonstrates that this certainly is the best of all possible worlds.

"It's certainly the most expensive," replies Allen.

Being mocked by Voltaire was certainly not the only knock that Leibniz took. For a century after he died, he was something of a pariah in England for his dispute with Newton and for his earlier opposition to John Locke, both national heroes.

Newton was the last man standing in the calculus wars, and he lived for another decade after Leibniz died. As an old man, he became a scientist of celebrity status in England and his fame spread abroad. Newton spent his autumn years constantly sought after by intellectuals and the well-to-do from England and abroad, who were excited to meet one of their heroes and one of the great minds of all time. Some of the scholars who visited him moved back to Europe, where they continued to champion his work.

Thus, Newton became more and more appreciated for his books, *Principia* and *Opticks*, in the last decade of his life, and he oversaw the publications of new editions of them. In the 1720s, his physics works were translated and lauded throughout Europe, and, in the decade after the calculus wars were cut short by Leibniz's death, his work in mathematics began to catch on outside of England.

It first happened in Holland. Even though England and Holland had fought more than one war in the seventeenth century, the rise of William of Orange to the throne of England had warmed relations dramatically. Besides, the Dutch were now free of the French and German bonds to Descartes and Leibniz, both of whom were threatened by Newton and his philosophy.

Hermann Boerhaave taught at Leiden, in Holland, and wholeheartedly embraced and disseminated Newton's philosophy. He called Newton the "Prince of Philosophers." Another advocate was Willem Jacob Gravesande, who has been called Newton's great Dutch popularizer. Gravesande also taught at Leiden—thanks in no small part to Newton, who had helped him obtain the position in 1717.

Even in France, with its long history of warfare and animosity with Great Britain, Newton was making headway—despite the fact that *Opticks* and *Principia* both were major challenges to aspects of Cartesian philosophy, and anti-Newtonianism had naturally arisen to counter the threat. The cooling of these tensions began in 1715, when an eclipse that was not visible in Paris but was in England brought a group of prominent intellectuals to London. Newton, as their gracious host, arranged for them to witness his optical experiments. He also saw that they were duly elected to the Royal Society. So full of gratitude was one member of the group, Pierre Rémond de Monmort, that he sent Newton fifty bottles of French champagne.

France began to warm to Newton after he was ultimately proven correct in one of his theories—namely that the earth is not a perfect sphere but an oblate spheroid that is flattened at the poles. In 1736, Pierre-Louis Moreau de Maupertuis went to Lapland to measure a minute of arc along the meridian. His careful survey proved Newton was right, and Maupertuis became Newton's champion in France—so much so that he was dubbed Sir Isaac Maupertuis.

By 1784, Newton's fame in France had grown so much that several competitions were held to design a monument in his honor. One of these was won by a man named Étienne-Louis Boullée, who designed a cenotaph—a tomb in which Newton's remains would not actually be held. It was a sphere several hundred feet high, with Newton's sarcophagus in the middle surrounded by a massive space. Another competition, held by the French Academy of Architecture the following year, called for design proposals "dedicated to the glory of the great genius, ought not to be magnificent so much as imposing in its dignified grandeur and noble simplicity."

After Newton died, he was the face of science, discovery, and other abstract notions of genius in the eighteenth century—much as Einstein was the face of genius in the twentieth—and his fame would continue to grow unabated. His image appeared in paintings, sculpture, and other art throughout the eighteenth century. Perhaps the most famous of all these statues is the one by Roubilliac that was erected on July 4, 1755, and now resides at Cambridge University. Newton is depicted standing on a pedestal in a loose gown, holding a prism and looking upward.

The well-to-do in Europe commissioned busts that they placed on their mantels or in other prominent places of display, and it became popular for people to have their portraits painted with such a bust in the background. Benjamin Franklin had one such portrait painted of himself.

The celebration of Newton appeared in literature as well as art. Joseph-Louis Lagrange, who is considered by some to be the greatest mathematician of the eighteenth century, called Newton the greatest and the luckiest of all mortals for what he accomplished. James Thomson wrote "A Poem Sacred to the Memory of Sir Isaac Newton" in which he referred to Newton as the all-piercing sage: "Shall the great soul of Newton quit this Earth/To mingle with the stars and every Muse/Astonish'd into silence, shun the weight/ Of honours due to this illustrious name." Voltaire put it simply, "Newton is the greatest man who has ever lived."

Even in recent years, the accolades continue to accumulate. An "Address from the Masters, Fellows, and Scholars of Trinity College to a Conference in Jerusalem Commemorating the 300th anniversary of the Birth of Isaac Newton" in February 1943 stated that "Homage to Newton is homage to the spirit of pure science." A few years ago, *Time* magazine named Newton the "man of the Seventeenth Century." And on September 12, 1999, *The Sunday Times* (London) named Newton the "Man of the Millennium," beating out other scientists such as Darwin and Einstein, as well as British politicians, poets, and patriots alike.

When Newton died, he left an estate valued at £32,000 that was willed to his closest living relatives, his half nephews and nieces from his mother's second marriage. More valuable than this sizeable fortune, however, was his reputation. He had become a living legend and was a highly sought-after London personality. By the time he died in 1727, he was at the absolute height of his fame, and dying was the only thing left for him to accomplish.

Death came to Newton shortly after he went to London at the end of February, to preside over his last Royal Society meeting on March 2. He looked great, and apparently felt great as well. He told his nephew-in-law, John Conduitt, that he had slept nine hours straight through a few days earlier.

However, on Friday, March 3, Newton became ill and returned home to rest. Unfortunately, he waited a week before contacting a doctor. On March 11, Conduitt heard that his uncle was ill, and he sent for a Dr. Mead and a Mr. Cheselden. These medical professionals diagnosed a stone in Newton's bladder, which probably caused Newton severe pain in his last few days. Despite the pain, he is said to have remained upbeat, and would smile while talking to visitors even as the beads of sweat rolled down his forehead. He seemed to recover slightly by the middle of the following week, and by Saturday, March 18, he was well enough to read the newspaper. Things were beginning to look as though he might survive the episode.

But by that night Newton was insensible, and he grew worse the next day, slowly succumbing over the course of many hours to his acute illness until he died at 1:00 a.m. on Monday, March 20, 1727. It was headline news in the British newspapers. One periodical declared Newton to be "the greatest philosophers and the glory of the English Nation." James Thomson quickly composed and published his "Poem Sacred to the Memory of Sir Isaac Newton," and, before the year was over, five separate editions of this poem had been published.

Compared to Leibniz's, Newton's funeral was an event for the ages. Newton had been larger than life, and he had a funeral worthy of such celebrity. He was interred in the nave at Westminster Abbey on March 28, 1726, where the kings and queens of England are crowned when they come to power and where they are buried when they die. Next to him lies the cream of the last several centuries of

British society—architects, scientists, poets, generals, theologians, and politicians—and he is buried among the likes of Dryden, Chaucer, Charles Darwin, Henry VIII, and Cecil Rhodes, and Mary, Queen of Scots.

His pallbearers were England's Lord Chancellor, the Dukes of Montrose and Roxburghe, and the Earls of Pembroke, Sussex, and Macclesfield. Along the procession, there were choirs and throngs of adoring masses paying their respects. The funeral mass itself was presided over by no less than a bishop.

Newton is buried beneath a marker on the floor of the nave—a big black stone that reads *Hic Depositum Est Quod Mortale Fuit Isaaci Newtoni* (The Mortal Remains of Isaac Newton). This stone is flanked by stones dedicated to the memories of Michael Faraday and James Clerk Maxwell—the highest company of British physicists.

An expensive monument was soon built in Westminster Abbey in Newton's honor, and the dean of Westminster found a very conspicuous place for it in the nave. Fatio assisted Conduitt with the design and inscription for the monument, and it was erected in 1731. It is a grand affair—a full-size statue of Newton at rest, reclined on a stack of books that represent what his contemporaries viewed as his major contributions to human knowledge when he died—his still-famous books on physics and optics, and his now-almost-forgotten contributions to theology and the chronology of ancient kingdoms.

To Newton's left are a couple of young angels displaying a diagram of the solar system. Above his head is a globe with a woman weeping atop it—Lady Astronomy, the queen of the sciences, in mourning. Beneath Newton rests a marble sarcophagus base with a relief work depicting either children or cherubs wielding the scientific tools of experiments that had made him famous: a reflecting telescope, a prism, a furnace, and a steelyard for weighing the planets, and money newly coined. One is decanting some liquid from one vial into another. Two youths stand before him with a scroll with a diagram of the solar system on it. Above that is a converging series.

The epitaph, translated, reads:

Here lies
Sir Isaac Newton, Knight,
Who, by a vigor of mind almost supernatural
First demonstrated
The motions and figures of the planets,
The paths of the comets and the tides of the ocean.
He diligently investigated
The different refrangibilities of the rays of light,
And the properties of the colors to which they rise.
An assiduous, sagacious, and faithful interpreter
Of nature, antiquity, and the holy scriptures,
He asserted in his philosophy the majesty of God
And exhibited in his conduct the simplicity of the gospel.
Let Mortals Rejoice
That there has existed such and so great
An ornament to the human race.
Born 25 December 1642 Died 20 March 1727

A 1726 portrait of a surprisingly young-looking Isaac Newton at the age of eighty-three portrays the distinguished scholar in his robes shortly before he died. He is depicted seated at a table with a copy of the newly printed third edition of his famous *Principia* open on his lap. The picture is inspiring—one of the greatest mathematicians of all time together with his greatest work. Newton is to mathematics and physics what Elvis Presley is to rock and roll—the icon who practically invented iconography. And Newton's *Principia*, his *opus magnum*, is a classic that ranks with Darwin's *Origin of the Species* as one of the most famous and most influential science books of all time. It continues to be translated from its original Latin even today.

The third edition of the *Principia* depicted in the painting is truly a handsome volume. I examined a copy at the Wren Library in Cambridge, and was impressed by its beauty. The fronispiece is a portrait print of Newton from 1725. This edition has more extensive tables of data than did previous volumes. It also includes a page with

Newton's name and an homage to the king—George II, George I's son, the second Hanoverian to rule England.

Newton rewrote this book throughout his entire life, and, through it and his other writings, he opened up whole new worlds of studies with his contributions to physics and optics, as well as invented the mathematical underpinnings needed to advance those disciplines. He developed mathematics as a way of rigorously describing physical phenomena—something that modern science takes almost for granted. Students of physics today may not ever read the *Principia*, but whether they know its text or not, the book has an indelible impact on their studies. Any student studying physics at the college level today will likely start the semester with a few weeks' worth of what is now called either classical mechanics or Newtonian mechanics.

And yet something is missing from this third edition. What is not in the picture hanging at the National Portrait Gallery is any indication of Newton's great rival Leibniz. Nor does the book that is open in front of Newton mention Leibniz's name. In the first edition of the *Principia*, Newton had acknowledged that Leibniz had invented his own form of calculus and that Leibniz's calculus had differed from his own only in notation and in the words they chose to describe this new branch of mathematics. That was in the 1680s. But for the second edition, which appeared in 1713, and for the edition of 1724, Newton had Leibniz edited out.

On display at the Leibnizhaus museum in Hanover, Germany, is a portrait of Leibniz that was painted prior to Newton's. Leibniz's shows him with a serious gaze and slightly furrowed brow. He has a bulbous nose, a slight double chin, a large head, and an even larger wig—a big, black, curly affair. One eyebrow looks ever so slightly raised, almost as if he is slightly amused. Or is he annoyed?

Leibniz left many things unfinished in his life—some, like the history of George I's family, were left to future generations to complete. When the books were finally released, it was not due to some overwhelming interest in the history itself but because of people's interest in publishing Leibniz's complete works. Other designs, ideas, and dreams of his will never be realized. He left a trail of these incomplete projects in his wake: the failed windmill project for the mines, advanced watches that he never built, his never-completed alphabet of human thought, new mechanical engines that never advanced beyond theory, and some swift carriages he dreamed up because the roads throughout Europe in his days were terrible

Ironically, despite all these unfinished projects, it was one of his most successful inventions, calculus, that would wind up defining failure for Leibniz. Had he been born at another time and accomplished the sorts of things he did without being under anyone else's shadow, he would be remembered now as the greatest mathematical and scientific mind of his day.

Leibniz was a mathematical novice who became, of his own volition, a math wizard. He was revolutionary for creating binary mathematics and advocating its use. He developed the use of determinants—a standard tool in linear algebra—and was of course revolutionary for both his invention and his dissemination of calculus. Indeed, he may have had one of the greatest minds of all time. He once boasted that he could recite almost all of Virgil's *Aeneid* by heart (one wonders if even Virgil could have done that). He was an accomplished lawyer and advisor whose skills were highly sought after. He was one of the most important philosophers of his day, a father of modern geology, and an expert on everything from biology and medicine to theology and statistics. A pen pal to scientists, diplomats, kings, queens, clergy, and medical doctors alike, he maintained lifelong correspondence with hundreds of his contemporaries on every imaginable subject.

He may have known as much about China as any European of his day—its history, technology, culture, religions, and even its flora, and fauna—and yet he never went there. All his information was obtained through books and by corresponding with Jesuit missionaries in China.

In short, aside from being an expert mathematician, he was a polymath—a man who not only had an interest in many different fields of knowledge, but who could contribute advances to these fields—and has been called a universal genius.

But in 1700, when he was generally regarded as the sole inventor of calculus and commanded the respect of most of the leading mathematicians in Europe, he took a mighty fall. Perhaps his fault was that he underestimated the threat that Newton's camp represented. He must have thought that he had truly invented calculus and had not borrowed anything from Newton, and that Newton himself recognized this fact. But in the years after Leibniz's death, there were probably few who would dispute that, at the very least, Newton was the first inventor of calculus, and many would buy Keill's argument that Leibniz may have indeed borrowed some of his calculus from Newton.

Did Leibniz lose the calculus wars?

In one sense, he did.

His life and legacy were both indelibly marked by the dispute, and even though he still had his supporters among the cadre of mathematicians he influenced and those mathematicians who followed them, that facet of his star faded after he died. He was never really able to promote his point of view concerning calculus's origins to the extent that popular opinion reverted back to where it had been two decades before he died—when, prior to any publication of Newton's mathematical discovery, Leibniz had been the unquestionable inventor of calculus.

Epilogue

n 1737, a few years after Newton died, his treatise *Method of Fluxions* finally appeared. This was the exposition of his method of calculus that he had written long before, and it was not printed as a bit of posthumous hero worship. The wording of the preface shows just how revered Newton had become only a decade after his death: "The following treatise containing the first principles of fluxions, though a posthumous work, yet being a genuine offspring (in an English dress) of the late Sir Isaac Newton, needs no other recommendation to the public than what that Great and Venerable Name will always carry with it."

Newton's discourse was at times hard to read. One striking example comes on page sixty, where he explains: "When a quantity is the greatest or the least that it can be at that moment it neither flows backwards nor forwards: for if it flows forwards or increases then it was less, and will presently be greater than it is; and on the contrary if it flows backwards or decreases, then it was greater and will be presently be less than it is. Wherefore to find its fluxion by [Newton's methods] and suppose it to be equal to nothing." The same meaning can be much more succinctly described today as: "set the derivative equal to zero and solve."

Nor was Newton's notation as useful as the superior notation that Leibniz had invented and the advanced calculus that Johann Bernoulli and the other European mathematicians developed throughout the century. Leibniz had correctly surmised that his symbols would make for the easy development of calculus, and these symbols, which he first penned in his notebooks in Paris in 1675, can still be found to this day in every calculus textbook.

In this sense, the high esteem in which Newton was held in Britain was not always a good thing, because, many of the mathematicians and scientists living there in the eighteenth century were behind the iron curtain of Newton's fame and glory. Ironically, as much as Leibniz's reputation suffered in Great Britain, the whole country may have suffered a self-inflicted wound by so underappreciating him. After the calculus wars, British mathematicians were prevented from learning calculus using Leibniz's notations, which were largely in use elsewhere, and they were not finally accepted in that country until the early nineteenth century.

It took until the mid-nineteenth century for explosion of scholarship to begin to redeem Leibniz and return to him the general recognition for his role in the creation of calculus. Even though he would no longer be regarded as its sole inventor, historians at that time would at least establish the facts that led to his now-universal regard as its co-inventor. It was their firm establishment of the basic facts of the calculus wars that led to this renewed appreciation of Leibniz's contributions. As one scholarly review of a new Leibniz biography in 1846 put it:

Most persons of the present day, who have investigated the subject, have pretty well made up their minds as to the following
points: first, that the system of Fluxions is essentially the same
with that of the Differential Calculus—differing only in notation; secondly, that Newton possessed the secret of Fluxions as early
as 1665—nineteen years before Leibniz *published* his discovery, and
eleven before he communicated it to Newton; thirdly, that both
Leibniz and Newton discovered their methods independently of
one another—and that, though the latter was the prior inventor
the former was also truly *an* inventor....Whether Leibniz was truly
an independent inventor of this method—in principle identical
with that of Fluxions—is the only question, in our judgment, that
really affects his fair name; and that he *was* so, is now, we may say,
all but universally regarded as indisputable.

Despite this writer's enthusiasm that the case was settled, some
scholars were still arguing even when he wrote these words. Some
nineteenth-century writers accepted Newton's stance that the sole
inventor was whoever had first come up with calculus and written
it down—thus giving himself full credit. After all, he did discover calculus first, twenty years before Leibniz published anything. To Newton, the discovery and subsequent dissemination of calculus were not
two parts of a whole discovery, and neither would they be to his subsequent champions.

To others, Leibniz was the one who deserved full credit, since his
methods and notation were the ones that progressed and survived.
He invented calculus independently, was the first to publish his
ideas, developed calculus more than had Newton, had far superior
notation, and worked for years to move calculus forward into a
mathematical framework that others could use as well. Besides, history is full of examples of second inventors taking full or partial credit
for an invention, including others from the seventeenth century.

Even so, in the mid-eighteenth century, many writers, like the
author of the review quoted above, began to take a more conciliatory

tone. In the century and a half since, some of Newton and Leibniz's biographers have gone even further and dismissed their fight as a ridiculous waste of time.

Actually, there is a long history of this sort of reasoning, dating all the way back to the middle of the calculus wars, as Varignon, a contemporary of the two mathematicians, first aired when he wrote a letter to Leibniz in 1713. Calculus was so great, Varignon said, that it should have been enough for both of them.

Another possibility is that neither one of them deserves all the credit that they were both seeking to claim from the other. In some ways, the development of calculus owes just as much all those who came before Leibniz and Newton, and to the Bernoulli brothers and the others like them who came afterward, took what was published, and turned it into a much richer subject with numerous applications.

For me, what's really interesting about the calculus wars is not who won or lost, but how they fought. The real story is not about how relevant or ridiculous the entire squabble was but how rich it was—and how much it reveals about both men.

Their stories were completely different. Leibniz went to Paris to avert a war and stayed to enrich his mind. He was embarrassed about his lack of knowledge in mathematics, but more than made up for it when he invented calculus, developed it, published it, and corresponded with others about it. While he was mired in his non-calculus-related obligations to the court at Hanover decades later, he was forced to defend his invention. Then, near the end of his life, he struggled in vain to beat down the accusations and insinuations that he was a plagarist. His story was tragic.

Newton's was triumphant. He invented calculus, wrote it down, shared it with a few people, forgot about it for a while, was asked about it, and again forgot about it for years. Then he began working on the *Principia* and, when he was finished, he found out that Leibniz had published own writings on calculus. For years, Newton held to the belief that he had been first to discover the process, and a few of his supporters came out and said as much in print, but he never did anything to win the glory of the invention for himself. Then, after a mid-life crisis, a new job at the mint, and a few years at the helm of the Royal Society, with the help of friends he launched a full-court effort to win recognition for his invention. And he ultimately succeeded.

Perhaps their argument reveals these men in their worst light. After all, theirs are two of the original profiles from which the archetypal myth of the modern scientist has been drawn—the ambitious, detached, hard working, prolific, and very nearly godlike genius—and one never likes to think of gods mired in nasty disputes. But then, perhaps the calculus wars reveal something more interesting.

It is a cautionary tale in the importance of publishing scientific discoveries, to be sure. Perhaps because Newton and Leibniz, fought the calculus wars at a time when each was at the height of his fame, the fight will forever be clouded in infamy to some. But to me it is one of the most fascinating stories in the history of science because it combines the most glorious heights of discovery with one of the most grueling and personal intrllectual fights. And it is possibly the only dispute in the history of science that was ever fought by two such great minds—perhaps the greatest of their day.

Bibliographical Essay

TWO SUMMERS AGO, when I was first starting to seriously work on this book, my wife and I were not yet married and were living what would turn out to be our last carefree summer before she became pregnant. One night we had an overnight guest at our place in the Bankers' Hill section of San Diego—an old friend from graduate school I hadn't seen in years. After a few beers he began asking me what I was working on, and I did my best to give him a synopsis . . . Newton, Leibniz and their famous fight.

My friend looked puzzled. "How do you become an expert in something like that?" he asked me. Though I was loathe to call myself an expert, my answer was basically good source material and an extraordinary amount of scholarship by generations of writers and academics who were interested in every aspect of their lives and work.

After Leibniz and Newton died, they both left large piles of papers, books that they had bought, and their correspondences, and these papers have been well preserved through the years because of their obvious importance as the life's work and thoughts of these two great men—from their boyhoods to their deathbeds and every stage in between.

This perception was especially true of Newton's papers, and since he was so famous in England his collection was instantly regarded as the treasure that it was. Ironically, as these papers were the embodiment of Newton's intellectual legacy, this legacy may have suffered somewhat because of his fame. Newton had carefully gone through and ordered his papers before he died, but in the years following his death, his legacy was shuffled, reshuffled, reordered, and finally divided.

At first, these papers became the property of John Conduitt, the husband of Newton's niece Catherine Barton, who was Newton's favorite relation. Shortly after Newton died, a Dr. Thomas Pellet was appointed to examine the papers and select those that were publishable. Almost none of them were, according to Pellet, and some of the papers today bear the legacy of this examination in the form off a note on their covers warning, "Not fit for publication." The only items in the entire mass that he selected for publication were short works on the chronology of the ancient kingdoms and a work called The System of the World, which Conduitt published soon after.

After Conduitt, the papers passed to he and Catharine Barton's son Lord Lymington, and from there they passed to a Mr. Saunderson in London, and eventually on to the Portsmouth family. Later, one of the Earls of Portsmouth allowed the university access to all the papers, which by this time, was not in the best of conditions. Some were water stained, others partially burnt, and many pages were not numbered and had fallen out of order. Besides that, some of the papers were on mixed subjects. There were theological papers, for instance, with mathematical notes in the margins. The decision then was to classify the papers into subjects like alchemy, chemistry, mathematics, chronology, history, and theology, and so the entire collection was reordered accordingly. Then it was split, and the earl donated those papers that related to mathematics to Cambridge University while keeping Newton's work on theology, the chronology of ancient kingdoms, and alchemy for himself.

From the nineteenth century onwards, Newton's biographers have more or less all been able to draw upon his papers and correspondence to aid in their work, and in the twentieth century, this primary source material became especially accessible with the publication of a set of seven volumes of Newton's correspondence was printed with notes and translations. The letters in this collection range from interesting historical texts to completely banal messages, such as the letter Newton wrote to Humfrey Ditton, March 16, 1714—right in the middle of the calculus wars. The letter reads, in its entirety: "Sir, If you please to call on me Friday morning next about ten of ye clock you will find me at home. I am Your most humble Servant Is. Newton." Other letters were much more valuable to me in the writing of this story, since they dealt directly with the calculus wars, and I have referred to these letters and in many cases quoted them directly throughout my book. Another useful work for a few of the early letters written by and about Newton was *The Correspondence of Henry Oldenburg, Volume IX.*

I should say that in many cases, I have taken the liberty of modernizing the spelling of certain words when I quoted from these letters. Words like "philosophicall," "concerne," "planetts," "centrall," and many more were altered to get rid of the extra vowels and consonants and others such as "ye" and "wch" were replaced with their obvious modernization. I also Americanized certain spellings like "favor." I'm sure some would bristle at the arbitrariness of my decision, but I felt that these spellings detracted rather than added anything, and so with "aapologies" to the editors of Newton's correspondence . . .

In addition to the seven volumes of Newton's correspondence, the *Principia* and *Opticks* are still in print and readily available. There are also numerous books, some of which can be found in the bibliography that follows, that excerpt passages and comment extensively on these texts. The most comprehensive and useful commentaries I found on Newton's great works were one by A.R. Hall on *Opticks* called *All Was Light* and an *Introduction to Newton's Principia* by I.B. Cohen.

These works are just the beginning. So much has been written about Newton, and so many times have his old writings and notes been gone through that there seems to be no end to Newtonian scholarship. People have read and printed and psychoanalyzed lists of words he wrote as a boy practicing his Latin grammar, and I once read a study by a top scholar looking at the way that books in his personal library were dog-eared—and what those dog ears reveal about his thought on important passages in books that he owned. And then there are the biographies—several of which I can mention.

The one biography I relied upon the most was *Never at Rest* by Richard Westfall, which was extraordinarily complete. Frank E. Manuel's *Portrait of Isaac Newton* was a very interesting read, particularly for his take on Newton's and I also liked the shorter and earlier work *Sir Isaac Newton* by Andrade. I found a useful book about Newton's time at the mint in Craig's *Newton at the Mint*. Of the older works, I enjoyed the great 1855 two volume *Memoirs of the Life, Writings, and Discoveries of Sir Isaac Newton* by Sir David Brewster. Another early book I read was Birch's *History of the Royal Society*, which provided some of the specific details for Chapter III.

There were a number of sources of information regarding Newton's growing fame and celebrity—the most obvious manifestation of which was his elaborate funeral, ornate tomb, and the explosion of poetry and art that invoked him. Some of the most interesting reading on Newton's influence of the worldview was written by Alexander Koyré, which I found in his *Newtonianism*. Koyré also expounds the clash between Newton's and Leibniz's metaphysics in an essay in the book by Frankfurt. This essay deals largely with the *Leibniz-Clarke Correspondence*, which is itself readily available in print in nicely translated and annotated editions.

Also helpful to me for understanding Newton's place in the world at the time that he died was the book by A.R. Hall called *Newton: Eighteenth Century Perspectives*, which contains some interesting biographies that appeared about him shortly after he died. Another book by Hall titled *The Revolution in Science 1500-1750* has a chapter devoted to Newton's Legacy, and yet another useful book that contains this sort of commentary is *Let Newton Be!*, which was edited by Fauvel et al.

A very visual presentation of Newton's influence, as seen in the art and writings of many in the eighteenth century, was the two-part show at the Huntington Gardens and Museum in Pasadena, California called *All Was Light*. This show, along with the companion book *The Newtonian Moment* by the show's curator Mordechai Feingold, were both very helpful to me because they presented copied of some of the original documents of the calculus wars, such as Newton's famous 1676 letters, and they focused on the growth and general acceptance of Newtonianism following his death.

Leibniz, too, left a pile of books, papers, and hand-written materials after he died, and because he spent his final years in the court library at Hanover, his collection of books and papers was naturally kept here as well. This created an interesting dilemma for King George and his family because Leibniz's papers were not solely important for their intellectual content. He had written numerous memoranda on subjects of courtly interest, political intrigue, and the goings-on at Hanover. As the new king of England, George was worried that these might shed a bad light on him or his family. When Leibniz died in 1716, George had barely been on the throne of England for two years, and the enemies to his reign were numerous. While Leibniz had been a loyal subject, his papers in the wrong hands might have provided some form of ammunition against George, so he took possession of everything.

This created a minor controversy as Leibniz's relatives had expected to inherit his books and papers. This was no insignificant inheritance—books were valuable items in those days, and Leibniz was famous so his papers were not without value either. The family took George to court, and the trial stretched on for years, decades, and was not decided for fifty years. Eventually the heirs were compensated for the value off the books, but the delay and the ultimate decision in the lawsuit meant that the pile of Leibniz's writing were kept essentially in one collection.

What a pile of papers it was. Leibniz left an overwhelming glut of papers, notes, and especially correspondence. By his own estimate, Leibniz wrote some 300 letters a year, which means that in the course of a decade, he would have written some 3,000, and over the five decades of his adult life, he would have written about 15,000—so much material, in fact, that by one estimate if a person sat down to read everything Leibniz ever wrote, assuming the read about 8 hours a day, it would take more than 20 years just to read all this writing— assuming, of course, that they could read the Latin, German, French, and the occasional Dutch, English, and that Leibniz corresponded in. "It would seem, indeed," one nineteenth-century biography put it, "as if these writings were a mine which could not be exhausted."

In today's world of email and text messaging, it may seem like a simple matter to send 300 letters in a single year—sometimes one might send three hundred emails in a single week. But there was a profound difference in what Leibniz was writing. Leibniz did not just dash off messages fit only for a chatroom ("LOL CU L8er loser") the way that people do today. Many of his letters were more like scholarly papers—the sort of which were fit for publication then and continue to be published today.

This is apparently not the easiest collection of papers to work with. Read Leibniz from the original sheets and you are not merely reading the words you are reading the scratch outs and the additions—all of which combine into a complicated fabric of a genius mind spilling forth, sometimes uncontrollably so. Copies of a few of his original letters are on display in the Leibnizhaus museum in Hanover, Germany. They are impressively detailed. His writing is tiny and exact, though in a script that was no doubt as thick as his accent was. In the tradition of the time, he writes over the entire surface of the page, sometimes writing additional comments vertically across the margins.

Perhaps because Leibniz's legacy was an unfinished encyclopedia instead of an *opus*, a large book for which he is primarily remembered the way that Newton is for the *Principia*, it was somewhat difficult to assemble a complete picture of his views. Some might argue that such a complete picture still does not exist anywhere, since despite nearly two centuries of intense study of his work, his complete works are still not published.

For years, a number of scholars have been undertaking the Herculean task of compiling the complete writings of Leibniz. The first attempts at this were made more than a century ago when a librarian in Hanover named G.H. Pertz tackled the history as his part. His colleague C.L. Grotefend helped him with the philosophical work, and C.I. Gerhardt helped him with the mathematical works. These mathematical works encompassed seven volumes, which were published in the mid-nineteenth century. And a few decades later Gerhardt contributed another seven volumes of philosophical works. Another eleven volumes of historical and political writings were produced by an O. Klopp, and an L.A. Foucher de Careil came out with seven volumes of history, politics, and church reunification.

Since this initial effort, a longer and more comprehensive effort has been underway to collect the complete works of Leibniz. The effort has been proceeding for several years without interruption in Germany at the library known as the Niedersachsische Landesbibliothek, a low-rise modern glass and concrete structure in the center of Hanover, which I visited during my research. Here and elsewhere scholars are collecting his letters, papers, and manuscripts into areas like law, politics, theology, history, philology, logic, geology, mathematics, and physics, and the effort underway

To date, more than half of what Leibniz wrote has been edited and published in some form or another, and as of March 2005, some 42 volumes of these writings had been collected in this definitive collection. Each volume comes in around 800 to 1,000 pages, and this is something less than half of the total. I have read that this work started in 1923, and one scholar estimates that when everything is finally collected, there may be perhaps 110 volumes in all. They are not yet halfway done with the effort, though it is estimated that they many reach the halfway point in the next decade.

Why was there so much writing? Leibniz traveled so extensively throughout Europe and maintained contact with the outside intellectual world through his extensive correspondence. He was willing to enter into correspondence with almost anyone. Many of these letters have been translated into English in individual books, which I purchased and read in the course of my research. Most notable were the translations by Leroy Loemker of several hundred pages of philosophical papers and letters. Also important for my efforts was a 1925 book entitled *Early Mathematical Manuscripts* by J.M. Child.

In addition to these "primary" sources, I drew heavily on a few biographies of Leibniz for this book. In the nineteenth century, there was an explosion of Leibnizian scholarship and a rediscovery of the value of his old papers and letters—at least a portion of them. A definitive biography, by a German scholar named Dr. G. E. Guhrauer, appeared in Germany in 1842 and drew heavily on the old papers. A biography based on Guhrauer's work and appeared in English in the mid-nineteenth century and was an enjoyable read for me. I refer to John Milton Mackie's *Life of Godfrey William von Leibnitz,* and this work provided many translations of Leibniz's letters from which I was able to pull quotes. Another mid-nineteenth century sketch that was useful to me was a review of Guhrauer's work that appeared in the *Edinburgh Review* in the mid-nineteenth century.

Worth noting as an aside is that there are many instances, especially in the older literature, in which Leibniz's name is spelled with a "t." Indeed, Newton, Keill, and many of Leibniz's contemporaries preferred to spell it that way, and the spelling persisted in English language works for more than a century after Leibniz died. In my book, I chose to use only the spelling without the t, and for the sake of avoiding confusion, I removed the alternative spelling where it appeared in the quotations of others.

A modern treatment of Leibniz's life can be found in Aiton's 1985 book *Leibniz* which is perhaps the best English language treatment of his life and works. Curiously, Aiton largely ignores the controversy over the invention of calculus, touching on it only incidentally. Nevertheless, without Aiton's thorough scholarship, it would not have been possible for me to penetrate the character of Leibniz nor assemble the facts that are represented in this book's narrative.

There were several other biographies that were also helpful to me.

Hofman's book *Leibniz in Paris* was a thorough and excellent examination of the years 1672-1676. Another interesting work, though much shorter, was Ross's *Leibniz*. Also helpful was a brief biosketch on Leibniz in Benson Mates's *The Philosophy of Leibniz* and a similar chapter in Jolley's *Cambridge Companion to Leibniz*.

In addition to these, there were a number of other books I read on Leibniz's work in other areas of that I only paid cursory attention to in my text. Leibniz's philosophy, political writings, and his writings on China, to name just a few areas, are rich and interesting, and while I read a few books on these subjects with interest, I was not able within the limited confines of my narrative to include everything—since my primary concern was the calculus wars.

The fight between Newton and Leibniz was so legendary that nearly every bio-sketch I found about either man touched on the calculus wars in part. And where some biographers, like Aiton, seem to consciously ignore the dispute, others, like Newton's biographer Westfall, devote considerable attention to it. To my knowledge, mine is the first book to tell the story of the calculus wars in a popular form, although Hall's *Philosophers at War* is an excellent scholarly history of the fight. For readers who want to know more about the details contained in this book, *Philosophers at War* is a great place to start.

Finally, suffice it to say that nobody can approach the writing of a story like this, which took place in the late seventeenth and early eighteenth centuries, without also becoming familiar with those times—the general political history of Europe in those days and the scientific revolution as a whole. I spent many afternoons perusing in the stacks of the central branch of the San Diego public library, and I have listed several books in the bibliography that helped me get more of a handle on those times. The books most useful for research on the House of Hanover were Redman's *The House of Hanover* and Black's *The Hanoverians*. Useful biographical information on some of the other mathematicians from the seventeenth century came from *A History of Mathematics* by Carl Boyer. Boyer's *History of Calculus* was also a useful read.

List of Illustrations

Bibliography

Ainsworth, John H., *Paper: The Fifth Wonder*. Wisconsin (1959).

Aiton, E. J., *Leibniz: A Biography*. Bristol (1985).

Alexander, H.G., ed., *The Leibniz-Clarke Correspondence*. Manchester (1998).

Algarotti, Sig., *Sir Isaac Newton's Philosophy Explain'd for the Use of the Ladies, Translated from the Italian*. Original edition in Wren Library, Cambridge (1739).

Andrade, E. N. da C., *Sir Isaac Newton*. London (1954).

Barber, W. H., *Leibniz in France from Arnauld to Voltaire: A Study in French Reactions to Leibnizianism, 1670–1760*. Oxford (1955).

Benecke, Gerhard, *Germany in the Thirty Years War*. New York (1979).

Bertoloni-Meli, Domenico, *Equivalence and Priority: Newton Versus Leibniz*. Oxford (2002).

Birch, T., *The History of the Royal Society of London for Improving Knowledge from its First Rise*. London (1756).

Black, Jeremy, *The Hanoverians: The History of a Dynasty*. London (2004).

Boyer, Carl, *A History of Mathematics, Second Edition*. New York (1991).

Boyer, Carl, *The History of the Calculus and its Conceptual Development (The Concept of the Calculus)*. New York (1959).

Brewster, Sir David, *Memoirs of the Life, Writings, and Discoveries of Sir Isaac Newton*. Edinburgh (1855).

Brown, Beatrice Curtis, *The Letters and Diplomatic Instructions of Queen Anne*. New York (1968).

Burrell, Sidney A., *Elements of Modern European History: The Main Strands of Development Since 1500*. Howard Chandler (1959).

Cairns, Trevor, *The Birth of Modern Europe*. Cambridge (1975).

Cajori, Florian, *A History of the Conceptions and Limits of Fluxions in Great Britian from Newton to Woodhouse*. Chicago (1919).

Cajori, Florian, "Leibniz, the Master Builder of Mathematical Notation" *Isis* 7, (1925), 412–429.

Cassirer, Ernst, "Newton and Leibniz." *The Philosophical Review, Volume 52*, 366–391 (1943).

Child, J.M., *The Early Mathematical Manuscripts of Leibniz*. Chicago (1920).

Clark, David, and Clark, Stephen P. H., *Newton's Tyranny: The Suppressed Scientific Discoveries of Stephen Gray and John Flamsteed*. New York (2001)

Cohen, I. Bernard, *Introduction to Newton's Principia*. Harvard (1999).

Cohen, I. B., and Westfall, R. S., ed., *Newton: A Norton Critical Edition*. New York (1995).

Cohen, I. B., "Newton's Copy of Leibniz's Theodicee: With Some Remarks on the Turned-Down Pages of Books in Newton's Library." *ISIS, 73*, 410–414 (1982).

Costabel, Pierre, *Leibniz and Dynamics*. Cornell (1973).

Craig, Sir John, *Newton at the Mint*. Cambridge (1946).

Davis, Martin, *The Universal Computer: The Road from Leibniz to Turing*. New York (2000).

Ditchburn, R. W., "Newton's Illness of 1692–3." *Notes and Records of the Royal Society of London, Volume 35*, 1–16, July (1980).

Durant, Will & Ariel, *The Age of Louis XIV*. New York (1963).

Durant, Will & Ariel, *The Age of Voltaire*. New York (1965).

Ede, Mary, *Arts and Society in England Under William and Mary*. London (1979).

Evans, R. J. W., "Learned Societies in Germany in the Seventeenth Century." *European Studies Review, 7*, 129–151 (1977).

Evelyn, John, *John Evelyn's Diary (Selections)*. Philip Francis, ed. London, (1965).

Fauvel, J., Flood, R., Shortland, M., and Wilson, R., *Let Newton Be! A New Perspective on his Life and Works*. Oxford (1988).

Feingold, Mordechai, *The Newtonian Moment*. New York/Oxford (2004).

Field, John, *Kingdom Power and Glory: A Historical Guide to Westminster Abbey*. London (2004).

Frankfurt, Harry G., ed., *Leibniz: A Collection of Critical Essays*. Notre Dame (1976).

Hall, A. Rupert, *All Was Light: An Introduction to Newton's Opticks*. Oxford (1995).

Hall, A. Rupert, *Isaac Newton: Adventurer in Thought*. Cambridge (1992).

Hall, A. Rupert, *Isaac Newton: Eighteenth Century Perspectives*. Oxford (1999).

Hall, A. Rupert, *Philosophers at War: The Quarrel Between Newton and Leibniz*. Cambridge (1980).

Hall, A. Rupert, *The Revolution in Science 1500–1750*. London (1989).

Hall, Marie Boas, *Nature and Nature's Laws*. New York (1970).

Hankins, Thomas L., "Eighteenth-Century Attempts to Resolve the *Vis viva* Controversy." *ISIS, 56*, 281–297 (1965).

Hofman, Joseph Ehrenfried, *Classical Mathematics: A Concise History of Mathematics in the Seventeenth and Eighteenth Centuries*. New York (1959).

Hofman, Joseph Ehrenfried, *Leibniz in Paris 1672–1676: His Growth to Mathematical Maturity*. Cambridge (1974).

Hollingdale, S. H., "Leibniz and the First Publication of the Calculus in 1684." *The Institute of Mathematics and its Application, Volume 21*, May/June (1985), 88–94.

Inwood, Stephen, *A History of London*. New York (1998).

Janiak, Andrew, ed., *Newton: Philosophical Writings*. Cambridge (2004).

Jolley, Nicholas, ed., *The Cambridge Companion to Leibniz*. Cambridge (1998).

Keynes, Milo, "Sir Isaac Newton and his Madness of 1692–93." *The Lancet*, March 8, (1980), 529–530.

Koyré, Alexandre, *From the Closed World to the Infinite Universe*. New York (1958).

Koyré, Alexandre, *Newtonian Studies*. Harvard (1965).

Koyré, Alexandre, and Cohen, I. Bernard, "Newton & the Leibniz-Clarke Correspondence with Notes on Newton, Conti & Des Maizeaux." *Archives Internationale d'Histoire des Sciences*, Volume 15, 63–126 (1962).

Langer, Herbert, *The Thirty Years' War*. New York (1978).

Leasor, James, *The Plague and the Fire*. New York (1961).

Leibniz, Gottfried Wilhelm, *Leibniz Selections*. Wiener, Philip P., ed. New York (1951).

Leibniz, Gottfried Wilhelm, *New Essays Concerning Human Understanding*. Alfred Gideon Langley, ed. Chicago (1994). Newport (1896).

Leibniz, Gottfried Wilhelm, *Writings on China*. D. Cook, and H. Rosemont, ed. Chicago (1994).

Lieb, Julian, and Hershman, Dorothy, "Isaac Newton: Mercury Poisoning or Manic Depression?" *The Lancet*, December 24/31 (1983), 1479–1480.

Loemker, Leroy E., *Gottfried Wilhelm Leibniz Philosophical Papers and Letters, Second Edition*. The Netherlands (1989).

Macaulay, Lord, *History of England*

Mackie, John Milton, *Life of Godfrey William von Leibnitz on the Basis of the German Work of Dr. G. E. Guhrauer*. Boston (1845).

Manuel, Frank E., *A Portrait of Isaac Newton*. Harvard (1968).

Mason, H. T., *The Leibniz-Arnauld Correspondence*, Manchester (1967).

Mates, Benson, *The Philosophy of Leibniz: Metaphysics & Language*. Oxford (1986).

Maury, Jean-Pierre, *Newton: The Father of Modern Astronomy*. New York (1992).

Moore, Cecil A., ed., *Restoration Literature: Poetry and Prose 1660–1700*, New York (1934).

Munck, Thomas, *Seventeenth Century Europe*. New York (1990).

Newton, Isaac, *The Correspondence of Isaac Newton, Volumes 1–7, 1661–1727*. Turnbull, Scott, Hall, and Tilling, ed. Caambridge, 1959–1977.

Newton, Isaac, *The Principia: Mathematical Principles of Natural Philosophy*. I.B. Cohen and Anne Whitman, ed. California (1999).

Newton, Sir Isaac, *Opticks or a Treatise of the Reflections, Refractions, Inflections, & Colours of Light* (Based on the Forth Edition). New York (1979).

Nussbaum, Frederick, *The Triumph of Science and Reason 1660–1685*. New York (1962).

Ogg, David, *England in the Reigns of James II and William III*. Oxford (1955).

Oldenburg, *The Correspondence of Henry Oldenburg Volume 9. 1672–1673*. Hall and Hall, ed. Wisconsin (1973).

Palter, Robert, ed., *The Annus Mirabilis of Sir Isaac Newton 1666–1966*. Cambridge, MA (1970).

Parker, Geoffrey, and Smith, Lesley, *The General Crisis of the Seventeenth Century*. New York (1978).

Parker, Geoffrey, "The 'Military Revolution,' 1560–1660—a Myth?" *Journal of Modern History 48*, 195–214, (1976).

Pepys, Samuel, *Passages from the Diary of Samuel Pepys*. Richard Le Gallienne, ed. New York (1923).

Ramati, Ayval, "Harmony at a Distance: Leibniz's Scientific Academies." *ISIS, 87*, 430–452 (1996).

Redman, Alvin, *The House of Hanover*. New York (1960).

Ross, G. MacDonald, *Leibniz*. Oxford (1986).

Ross, G. MacDonald, "Leibniz and the Nuremburg Alchemical Society." *Studia Leibnitania, Band VI, Heft 2* (1974).

Rowen, Herbert H., *A History of Early Modern Europe 1500–1815*. New York (1960).

Russell, Bertrand, *A Critical Exposition to the Philosophy of Leibniz*. London (1937).

Rutherford, Donald, "Demonstration and Reconciliation: The Eclipse of the Geometrical Method in Leibniz's Philopsophy." *Firenze*, (1996), Leo S. Olschki, ed. 181–201.

Rutherford, Donald, "Leibniz: I Volume 12 E 13. Dell'Edizione Dell'Accademia." *Il Cannocchiale Rivista Di Studi Filosofici N. 3*, Settembre-Dicembre (1992), 69–75.

Scriba, Christoph J., "The Inverse Method of Tangents: A Dialogue between Leibniz and Newton (1675–1677)." *Archive for History of Exact Sciences, Volume 2*, (1964).

Symcox, Geoffrey, *War, Diplomacy, and Imperialism, 1618–1783*. New York (1974).

Voisé, Waldemar, "Leibniz's Model of Political Thinking." *Organon 4*, 187–205 (1967).

Voltaire, *Ancient and Modern History, Volume Six*. New York (1901).

Voltaire, *Candide*. New York (1930).

Voltaire, *Letters Concerning the English Nation*. Nicholas Cronk, ed. Oxford (1999).

Westfall, Richard, *Never at Rest: A Biography of Isaac Newton*. Cambridge (1980).

Westlake, H.F., *The Story of Westminster Abbey*. London (1924).

White, Michael, *Isaac Newton The Last Sorcerer*. London (1998).

Whiteside, D. T., *The Mathematical Papers of Isaac Newton, Volumes I-VIII*. Cambridge (1968–1981).

Whiteside, D.T.,The Mathematical Principles Underlying Newton's *Principia Mathematica*. *Journal for the History of Astronomy*, i, 116–138 (1970).

OTHER WORKS CITED

Address to the Masters, Fellows, and Scholars of Trinity College to a Conference in Jerusalem Commemorating the 300th Anniversary of the Birth of Isaac Newton. Original edition in Wren Library, Cambridge, dated February, 1943.

Calculating Machine. A display at the Niedersaechsische Landesbibliothek, Hanover Germany.

A Catalogue of the Portsmouth Collection of Books and Papers Written by or Belonging to Sir Isaac Newton. Original edition in Wren Library, Cambridge. Cambridge (1888).

Commercium Epistolicum. A 1722 copy that exists in the Royal Society Library, London England.

Communication Made to the Cambridge Antiquarian Society No. XII. Cambridge (1892).

Leibniz Korrespondenz. Display at Leibnizhaus, Hanover, Germany.

Leibniz Reisen. Display at Leibnizhaus, Hanover, Germany.

Lowery, H., "Newton Tercentenary, 1642–1942." An original copy at the Royal Society of London reprinted from the *Dioptric Review and the British Journal of Physiological Optics, Volume 3*, 105–113.

Newton, Sir Isaac, *A Treatise of the Method of Fluxions and Infinite Series with its Application to the Geometry of Curve Lines*. Original edition in Wren Library, Cambridge (1737).

"Gottfried Wilhelm Freiherr von Leibnitz—Eine Biographie (Review)." *The Edinburgh Review, Volume LXXXIV, Number CLXIX*, July, (1846).

The Wren Library Trinity College Cambridge. Informational pamphlet dated April (2004).

Index

Acknowledgments

Writing a book of twelve chapters is a lot harder than writing twelve separate essays, a more experienced writer once warned me when I was first starting this book a couple of years ago. Now, on the other end of things, I have to say she was right—right in the same way that being married is harder than dating or that raising a kid is a lot harder than having a pet.

Now that I am nearly finished, my friends and relatives have been asking me how I feel, and I keep saying the same thing. In the last two years I have gotten engaged, nursed my fiancée back to health from a bad collar bone injury, got married, watched the birth of our daughter, Georgia, quit my job, moved from the west coast to the east coast, started a new job, and renovated a home. All the while, in the background and in the foreground was this book, these twelve chapters, and what I feel now most of all is gratitude. I could not have done this without help from numerous people.

First off, thanks to my agent Giles Anderson with whom I began this book some three years ago. He was a tireless advocate throughout the project. Thanks, too, to Jenny Meyer and her agency for their efforts in conjunction with Giles.

Special thanks to my publisher John Oakes and all the folks at Thunder's Mouth Press and the Avalon Publishing Group who saw the book through to completion. Thanks to Jofie Ferrari-Adler, who was the initial editor of the book and Iris Bass who edited my complete manuscript. Special thanks to publicist Anne Sullivan, and thanks, too, to John's editorial assistant Lukas Volger.

I would like to thank Herbert Breger at the Gottfried Wilhelm Leibniz Bibliothek in the Niedersaechsische Landesbibliothek in Hanover, Germany for meeting with me and discussing Leibniz for several hours on one cold January afternoon. Dr. Breger also very graciously helped me obtain a few of the images that are published in this book by putting me in touch with his colleagues Dr. Friedrich Hülsmann and Birgit Zimny at the Gottfried Wilhelm Leibniz Bibliothek. Special thanks to Dr. Hülsmann and Birgit.

For the other images, I have to extend most gracious thanks to Christine Woollett at the Royal Society. Thanks also to Christine Falcombello of The Centre Iconographique Genevois and to the staff of the Library of Congress Prints and Photographs Division.

Special thanks to Kerstin Hellmuth of the University of Hanover for her tour of Leibnizhaus. Thanks too to Donald Rutherford for the conversations we had at a coffee shop at the University of California, San Diego.

Thanks to Richard Crawford, the librarian at the Wangenheim room of the San Diego Public Library's Central Library, and also to the rest of the staff at the library for their various assistance in interlibrary loans and other tasks. I also made use of and would like to acknowledge the Vancouver Public Library; the Geisel Library at the University of California, San Diego; the Library at the Royal Society in London; and the Wren Library at Trinity College, Cambridge. Special thanks to David McKitterick and Joanna Ball for access to the collections at the Wren Library. Also to Nigel Unwin, who helped me gain access to the library and to Ian and Jennifer Glynn, who had me over for a nice cup of tea and conversation after a long day of browsing 300-year-old manuscripts at the Wren.

Also special thanks to Mika Benedyk for her initial read of my chapters and many, many useful comments. Thanks to Kevin Fung for designing my web site and flyers. Special thanks to my friends and coworkers at The Scripps Research Institute for all the useful

discussions along the way—especially to Tamas Bartfai, who asked me about the progress of my book on several occasions in the last couple of years. Thanks, too, to Keith McKeown, who was supportive all along the way.

I would also like to thank the numerous members of my extended family who were always there for me when it counted. Lucy, Al, JB, Amy, Ty, Gina, Bruce, Jayne, Barb, Alf, Brian, Michelle, Andy, Ariel, Abby, Julian, and Jeannette. And thanks to all my friends with whom I have discussed this material on many occasions: Johan, Paula, John, Mr. Chart, Ellen, Nick, and Teddy.

Most of all, I would like to thank my wife Jennifer and our beautiful baby Georgia, without whom none of this would have been possible. Even as I write this acknowledgement, one of my last acts in the drafting of this book, I am listening to them play in the next room. Jennifer is singing and blowing kisses on her little belly. Georgia is laughing and chirping "Digoy Digoy Digoy!" Theirs is a love that makes me believe that this might very well be the best of all possible worlds.

117, 1684 cover letter mentions Newton

235 Montmort + Newton 1715